Welcome to
Teach yourself
Lightroom

Welcome to the latest edition of our Lightroom bookazine! Adobe's excellent software is a dual-purpose tool. It organises your images and enhances them too, so we start out here by looking at its import and cataloguing tools. Lightroom's editing tools grab the headlines, but it's often in the Library module that the really important work is done as you learn to sort, search and organise an ever-growing collection of photos.

Next, we run through the image enhancement tools in Lightroom's Develop module, from basics such as cropping and straightening, to tone curve adjustments and the increasingly powerful localised adjustment tools. We also run through Lightroom's Print, Map and Web modules, as it's no use collecting a vast library of photos if you don't do anything with them. And we've a specially expanded section on Lightroom's increasingly important mobile app.

We hope you enjoy our new, updated guide to Lightroom and discover how it can help you to transform your own photos into images you can be proud of.

FUTURE

Teach yourself Lightroom

Future PLC Quay House, The Ambury, Bath, BA1 1UA

Editorial
Editor **Rod Lawton**
Designer **Rodney Dive**
Compiled by **Philippa Grafton & Briony Duguid**
Head of Art & Design **Greg Whitaker**
Editorial Director **Jon White**
Managing Director **Grainne McKenna**

Photography
All copyrights and trademarks are recognised and respected

Advertising
Media packs are available on request
Commercial Director **Clare Dove**

International
Head of Print Licensing **Rachel Shaw**
licensing@futurenet.com
www.futurecontenthub.com

Circulation
Head of Newstrade **Tim Mathers**

Production
Head of Production **Mark Constance**
Production Project Manager **Matthew Eglinton**
Advertising Production Manager **Joanne Crosby**
Digital Editions Controller **Jason Hudson**
Production Managers **Keely Miller, Nola Cokely, Vivienne Calvert, Fran Twentyman**

Printed in the UK

Distributed by Marketforce – www.marketforce.co.uk
For enquiries, please email: mfcommunications@futurenet.com

Teach Yourself Lightroom Eleventh Edition (PTB6410)
© 2024 Future Publishing Limited

We are committed to only using magazine paper which is derived from responsibly managed, certified forestry and chlorine-free manufacture. The paper in this bookazine was sourced and produced from sustainable managed forests, conforming to strict environmental and socioeconomic standards.

All contents © 2024 Future Publishing Limited or published under licence. All rights reserved. No part of this magazine may be used, stored, transmitted or reproduced in any way without the prior written permission of the publisher. Future Publishing Limited (company number 2008885) is registered in England and Wales. Registered office: Quay House, The Ambury, Bath BA1 1UA. All information contained in this publication is for information only and is, as far as we are aware, correct at the time of going to press. Future cannot accept any responsibility for errors or inaccuracies in such information. You are advised to contact manufacturers and retailers directly with regard to the price of products/services referred to in this publication. Apps and websites mentioned in this publication are not under our control. We are not responsible for their contents or any other changes or updates to them. This magazine is fully independent and not affiliated in any way with the companies mentioned herein.

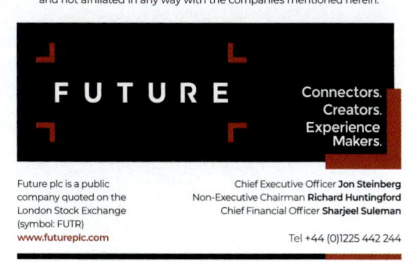

FUTURE Connectors. Creators. Experience Makers.

Future plc is a public company quoted on the London Stock Exchange (symbol: FUTR)
www.futureplc.com

Chief Executive Officer **Jon Steinberg**
Non-Executive Chairman **Richard Huntingford**
Chief Financial Officer **Sharjeel Suleman**

Tel +44 (0)1225 442 244

Teach yourself **Lightroom**

CONTENTS

CHAPTER 1

Import and sort images   8

Introducing the Lightroom workspace 10
Import your photos into Lightroom 12
Find photos fast using embedded metadata 14
Organise images using Collections 16

CHAPTER 2

The Library module   18

Introducing the Library module 20
Sort and rate your images in Lightroom 22
Find images fast with keywords 24
Tag your images with location data 26

CHAPTER 3

The Develop module   28

Introducing the Develop module 30
Discover how to use the histogram 32
Use Smart Previews to edit remote images 34
Make basic colour adjustments 38
Crop and straighten your photographs 40

CHAPTER 4

Photo-fixing tools   44

Remove distortion caused by your lens 46
Get a better perspective in your images 48
Reveal more midtone detail and texture 50

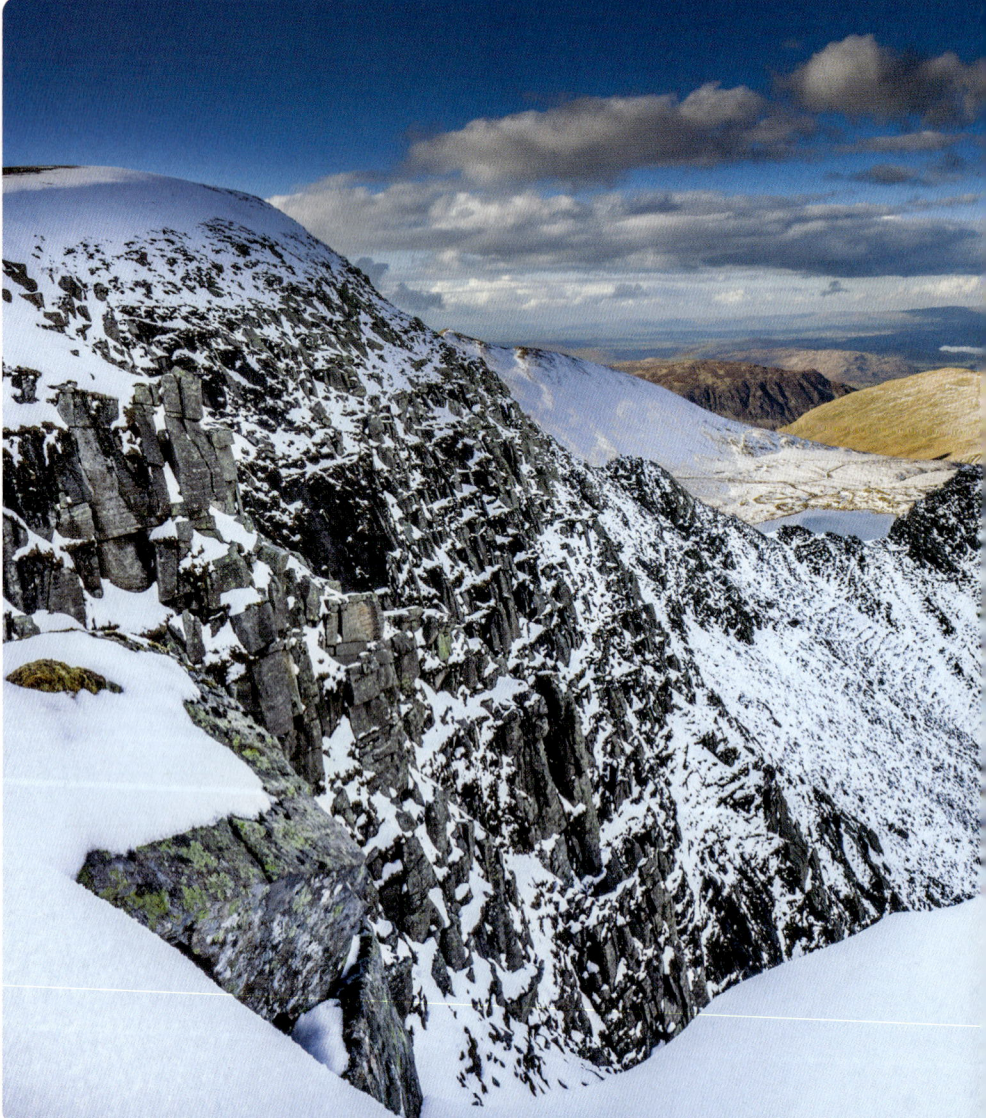

Teach yourself Lightroom

CONTENTS

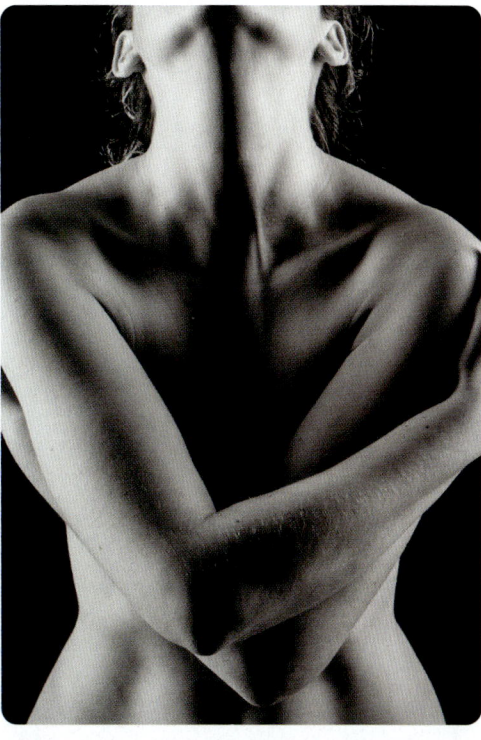

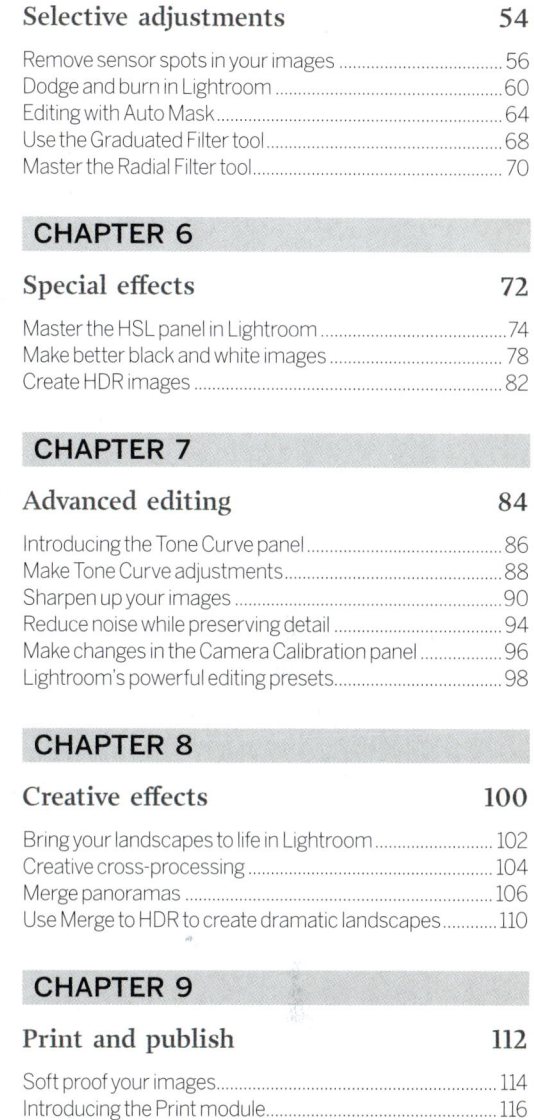

CHAPTER 5
Selective adjustments   54
- Remove sensor spots in your images 56
- Dodge and burn in Lightroom 60
- Editing with Auto Mask 64
- Use the Graduated Filter tool 68
- Master the Radial Filter tool 70

CHAPTER 6
Special effects   72
- Master the HSL panel in Lightroom 74
- Make better black and white images 78
- Create HDR images 82

CHAPTER 7
Advanced editing   84
- Introducing the Tone Curve panel 86
- Make Tone Curve adjustments 88
- Sharpen up your images 90
- Reduce noise while preserving detail 94
- Make changes in the Camera Calibration panel 96
- Lightroom's powerful editing presets 98

CHAPTER 8
Creative effects   100
- Bring your landscapes to life in Lightroom 102
- Creative cross-processing 104
- Merge panoramas 106
- Use Merge to HDR to create dramatic landscapes 110

CHAPTER 9
Print and publish   112
- Soft proof your images 114
- Introducing the Print module 116
- Create a custom print layout 118
- Watermark your images 120
- Publish your pictures with Lightroom Web 122
- Create an online photo portfolio 124

CHAPTER 10
Advanced skills   128
- Get creative with Lightroom 130
- Retouch your images like a pro 138

CHAPTER 11
Lightroom Mobile   144
- Sync Lightroom Mobile 146
- Working with Lightroom Mobile 148
- Sort your Library with Lightroom Mobile 150
- Shooting super images with Lightroom Mobile 152
- Lightroom Mobile's powerful editing tools 156

Teach yourself **Lightroom**
IMPORT AND SORT IMAGES

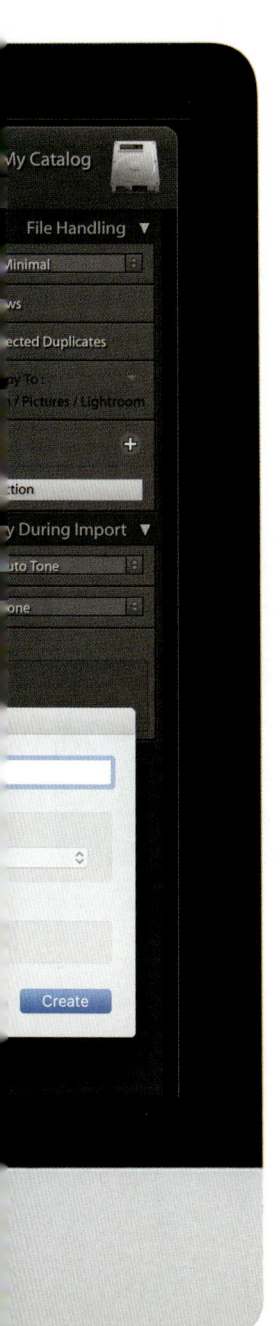

Teach yourself **Lightroom**

IMPORT AND SORT IMAGES

Import and sort images

Get started in Lightroom fast by learning the best ways to import and organise your images

10 Introducing the Lightroom workspace
Get to know the general layout and purpose of Lightroom's photo organising, editing and sharing modules

12 Import your photos into Lightroom
Discover how to import your photos and video clips from a memory card, camera or hard drive into Lightroom

14 Find photos fast using embedded metadata
Protect your images by assigning your copyright and adding your contact details to every picture's metadata

16 Organise images using Collections
Learn how to sort your photos into themed Collections, and use Smart Collections to automatically file your shots for you

Teach yourself Lightroom

IMPORT AND SORT IMAGES

GET THE FILES HERE: http://bit.ly/tylr2016

The Lightroom workspace

Get to know the general layout and purpose of Lightroom's photo organising, editing and sharing modules

Photoshop Lightroom combines the professional raw-processing tools in the more expensive Photoshop CC with the asset-organising powers of the cheaper Photoshop Elements. It also has plenty of unique photo-fixing and organising tools of its own, as you'll discover while working through this book.

One of the biggest challenges that we face as digital photographers is managing our collections of images. Lightroom enables you to take the tedium out of asset management. It provides easy ways to add keywords to batches of files as you import them from a memory card, so you can find specific images more quickly in the future. We'll look at Lightroom's asset management functions in more detail in chapter two.

You may be dealing with thousands of images that are scattered across folders on your PC as well as on external hard drives. Lightroom enables you to gather stray photos into its Catalog regardless of their locations, and organise them in a variety of ways. You can create themed collections, or even set up Smart Collections, which automatically collect files that meet specific criteria — more on this in a few pages. But first we'll give you an overview of Lightroom's layout.

Lightroom has seven key workspaces, called modules. The main modules are Library and Develop, because these enable you to import, organise and process your photos. The other modules are Map, Book, Slideshow, Print and Web. Check out our annotated grab to familiarise yourself with the Module picker.

Teach yourself **Lightroom**

IMPORT AND SORT IMAGES

Lightroom Anatomy Module picker

Get to know the seven key modules in Lightroom's workspace

1 LIBRARY

Lightroom opens in the Library module. The Library module is where you import and organise your digital photos. You can also tweak their colours and tones using the Quick Develop panel.

2 DEVELOP

The Develop module is your digital darkroom. Here you can fix problems relating to colour, tone and composition. This module also has a collection of selective raw editing tools.

3 MAP

If your camera has the capacity to geo-tag photos with GPS coordinates you can see images on a map according to where they were captured. You can also use Lightroom to manually geo-tag photos.

4 BOOK

Once you've processed your pictures, you may want to share them in a photo book. This module enables you to lay out the book using handy templates, and export the book to a publisher who will produce a hard copy.

5 SLIDESHOW

In this module you can present your photos as a slick slide show. In chapter ten we'll look at this module in more detail, and show you how to add music and transitions to your show, and how to watermark your images to protect them.

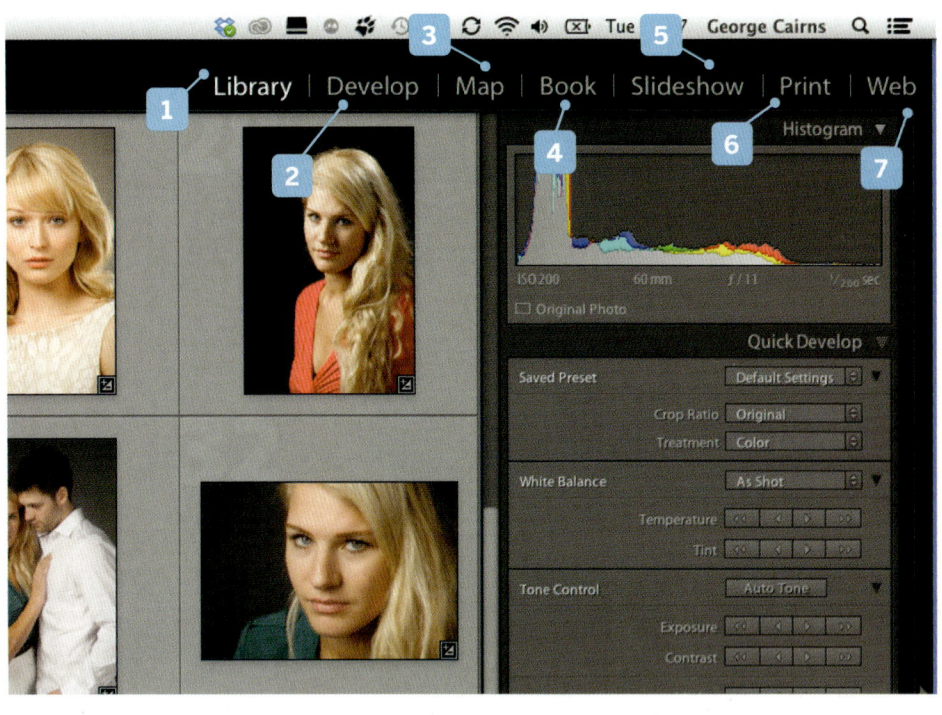

6 PRINT

In chapter nine we'll show you how to produce perfect prints using the tools and templates in this module. This chapter will also demonstrate how to make sure that what you see on screen looks the same in print using Lightroom's soft-proofing tools.

7 WEB

This handy module provides you with a host of templates so that you can share your photos online in interactive and attractive web galleries. Lightroom enables you to make web galleries without any coding knowledge at all.

Understanding...
PANEL CONTROL IN LIGHTROOM

Each Lightroom module has a range of panels, such as the Library module's Keywording panel [1].

Some panels may be more useful to you than others, so you can minimise the ones you want to hide by clicking these icons [2]. The hidden panels will reappear when you move the cursor near the sides (or top and bottom) of the screen. If you want less clutter and to get rid of those displays that you don't use, then right-click a panel to activate a context-sensitive pop-up menu.

Clear any panels [3] that you don't often use to make more space. The Solo Mode [4] option means that you have one open panel at a time.

KEYBOARD SHORTCUTS

With seven modules to explore, it's worth mastering the keyboard shortcuts that can summon each module with a few taps. The seven modules can be summoned with sensibly-numbered keyboard shortcuts that relate to the order in which they appear at the top of the workspace.

Use Cmd/Ctrl+Alt+1 to access the Library module, Cmd/Ctrl+Alt+2 for the Develop module, Cmd/Ctrl+Alt+3 for Map, Cmd/Ctrl+Alt+4 for Book, Cmd/Ctrl+Alt+5 for Slideshow, Cmd/Ctrl+Alt+6 for Print, and Cmd/Ctrl+Alt+7 for Web.

When you're using Lightroom, you'll spend most of your time in the Develop module, so you can also jump straight there with a snappier tap of the D key.

Teach yourself Lightroom

IMPORT AND SORT IMAGES

GET THE FILES HERE: http://bit.ly/tylr2016

Import your photos into Lightroom

Discover how to import images and video clips from a memory card or hard drive into Lightroom

1 Choose a source

Lightroom collects files from a variety of sources and displays them in its Catalog. After launching Lightroom, click the Import button at the bottom-left of the interface (or choose File›Import Photos and Video from the main menu.) An import window will appear. In the Source section, browse the files and folders and choose a source such as a memory card, your camera if it's plugged in, or a folder of photos on an internal or external hard drive.

2 Check or uncheck?

The assets in the selected source folder or memory card will appear as thumbnails. Use the slider to increase the thumbnail size for a closer look. All the files are checked automatically. You can tick Uncheck All and then manually check the thumbnails of all the photos that you want to import. The unchecked thumbnails will have a vignette around their edges. If you hold down Alt, the buttons will change to enable you to check or uncheck any video files.

Teach yourself Lightroom
IMPORT AND SORT IMAGES

3 Copy or Add?
If you're importing files from a memory card, click Copy. This will copy them to a chosen location on your computer's hard drive (such as your Pictures folder), or onto an external hard drive. It will then import the copied photos into Lightroom's Catalog. If you're importing files from a folder on an internal or external hard drive, click Add. This will add them to Lightroom's Catalog without physically moving or copying them, which saves disc space.

4 Loupe or Grid view?
By default you will see the targeted folder's content as thumbnails in the Grid view. To see a particular photo more clearly, either click the Loupe view icon or double click the thumbnail to take it into the Loupe view. You can then tick the Include in Import box if you want to add the photo to Lightroom's Catalog. Press G to go back to Grid view, where you can see the folder's contents as thumbnails.

5 Process while importing
You can batch process files as you import them. By checking the Build Smart Previews box you can edit copies of a photo even if the external hard drive that it's stored on has been disconnected. You can also apply preset Develop Settings to every photo, which enables you to creatively process them or apply automatic tonal corrections as they are imported. We'll look at ways to batch process imported files in more detail later in this chapter.

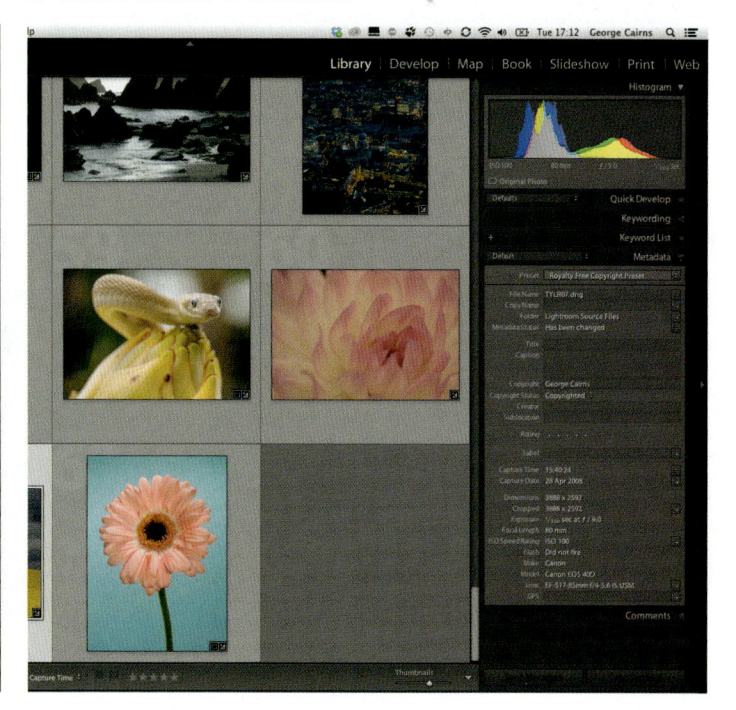

6 Import the images
Once you've decided which photos you want to include in Lightroom's Catalog, and you've chosen any File Handling or Develop Settings presets, click Import. A progress bar will indicate that the files are being imported into the Catalog from their current location. Lightroom will then generate standard previews of each file. In our example we chose not to apply any Smart Preview or Develop Settings presets, because we'll look at these options later.

Teach yourself Lightroom
IMPORT AND SORT IMAGES

GET THE FILES HERE: http://bit.ly/tylr2016

Find photos fast using metadata

Protect your images by assigning your copyright and adding your contact details to every picture's metadata

It's common practice to share your digital photos as electronic copies. You might do so by emailing them to clients, putting them on your social network site or presenting them in your online gallery. Once you've shared a photo you can't control where it ends up, because it's easy for others to make an electronic copy. This can lead to scenarios where your work is shared or published without you being credited or paid for it. As the creator of the image, you own the copyright to it, so others must seek your permission to use it. To help them do so you can assign your copyright details to the photo's metadata.

When your camera processes an image to describe its colours and tones, it also includes information about the camera settings used to capture the photo, such as the shutter speed and aperture setting. This metadata is stored with the image file, so that wherever the image goes, the metadata goes too. By inserting your copyright and contact details into the metadata, your photo is less likely to be used without you being credited.

Lightroom's Metadata panel is designed to enable you to add important information such as whether the photo is copyrighted and who created it. We'll show you how to edit a photo's metadata and then turn that information into a metadata preset, so that you can apply your copyright details to multiple photos with a few clicks. We'll also show you how to quickly apply presets to multiple photos as you import them, which is a huge time-saving step in your picture processing workflow.

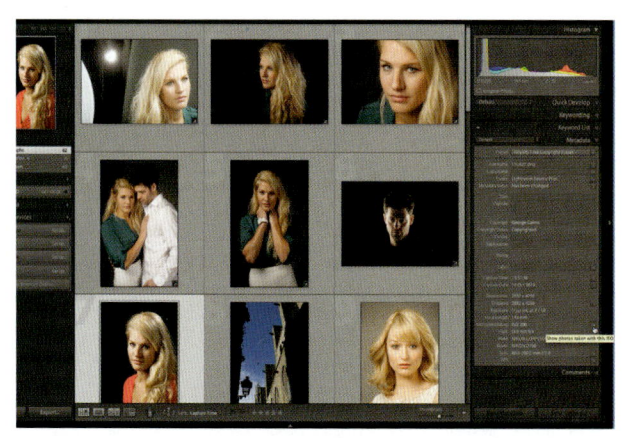

Teach yourself Lightroom

IMPORT AND SORT IMAGES

1 Filter using metadata
In the Library module, click a thumbnail. In the right-hand panel, toggle open the Metadata tab. Here you'll see information such as the Capture Date. If you click the arrow to the right of the ISO Speed Rating you can display all the other photos captured using the same setting.

2 Add the copyright status
Details such as ISO Speed rating and the Lens used can't be altered, but there are blank fields that you can type in to add to the selected photo's metadata. Type your name into the photo's Copyright field. Set the Copyright Status drop-down menu to Copyrighted (or Public Domain).

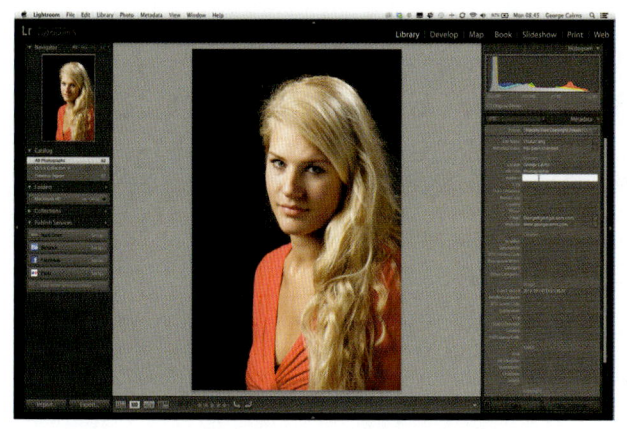

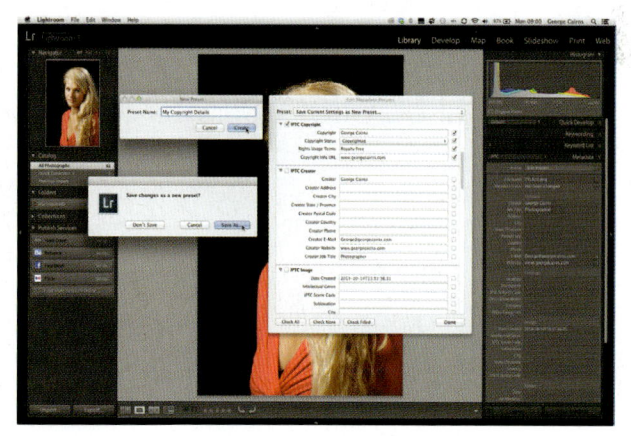

3 Add more personal details
To add more detailed information, click the Default drop-down menu to the left of the Metadata panel's label and choose IPTC. This gives you more text fields to edit, so you can add your address and website details. This will help others contact you should they wish to use your image.

4 Make a metadata preset
To speed things up, click the Metadata panel's Preset drop-down menu and choose Edit Preset. Continue editing the fields and choose Save Current Settings as New Preset from the Edit Metadata Preset window's drop-down. Click Done. Click Save As. Name the Preset. Click Create.

5 Assign multiple presets
Hold down Shift and click on images to select multiple thumbnails in Lightroom's Grid view. Go to the Preset drop-down menu on the Metadata tab and choose the custom preset you created in the previous step. Any details you included will be added to the selected photos.

6 Assign presets on import
You can assign your copyright details to every photo as you import it. Click Import and browse to the folder or memory card that you want to import. In the Apply During Import Panel, open the Metadata drop-down menu and choose the desired custom preset.

Teach yourself **Lightroom**
IMPORT AND SORT IMAGES

GET THE FILES HERE: http://bit.ly/tylr2016

Organise images using Collections

Learn how to sort your photos into themed Collections, and use Smart Collections to automatically file your shots for you

Thanks to digital cameras, you can generate thousands of photographs in a relatively short time. Lightroom gathers images from multiple folders and external hard drives and stores a link to them in its Catalog. You might have to spend a long time scrolling through the Library module's imported thumbnail images in search of a specific photo. The Catalog panel has the option to display all the photos in the Catalog, or you can narrow things down by clicking the Previous Import label to see your most recent additions to the Catalog.

In chapter two we'll look in detail at the ways Lightroom's Library module enables you to organise your images using a range of tools and commands. However, we'll pre-empt that chapter by introducing you to a quick and effective way to begin organising and displaying your imported photos.

In bygone analogue days we'd gather our favourite prints into albums so that we could browse through the images. Albums tended to collect and present our photos according to themes (such as a wedding or a holiday). This traditional model of storing and accessing images can be applied to the photographs in the Lightroom Library module, courtesy of the Collection panel.

In this walkthrough we'll show you quick ways to gather specific photos into Collections, so that you can display a small Collection in the Grid view instead of having to scroll through thousands of thumbnails in the main Catalog. You can create as many Collections as you like. We'll also demonstrate how to get Lightroom to automatically create Collections according to specific metadata criteria, using the aptly named Smart Collections tab.

1 Make a Quick Collection
The Catalog panel on the left of the Library module has a Quick Collection label. To quickly add our supplied landscape-themed photos to a Quick Collection, click each of the thumbnails and press B. Here we've collected six landscape-themed photos together in a Quick Collection.

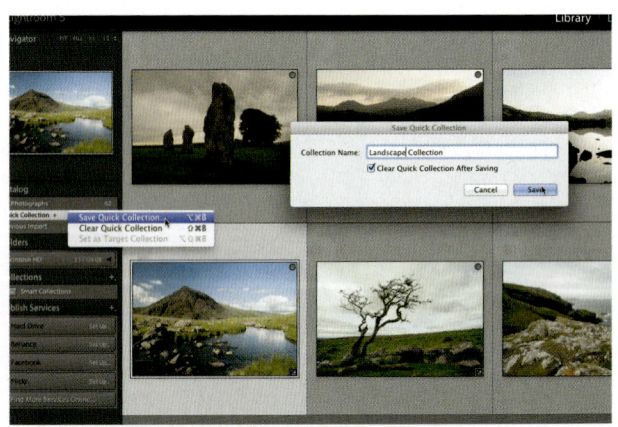

2 Save the Quick Collection
Go to the Catalog panel and click the Quick Collection label. The Grid view will now display the images you gathered. Right-click the Quick Collection label to summon the Save Quick Collection window. Label it 'Landscape Collection'. Click Save.

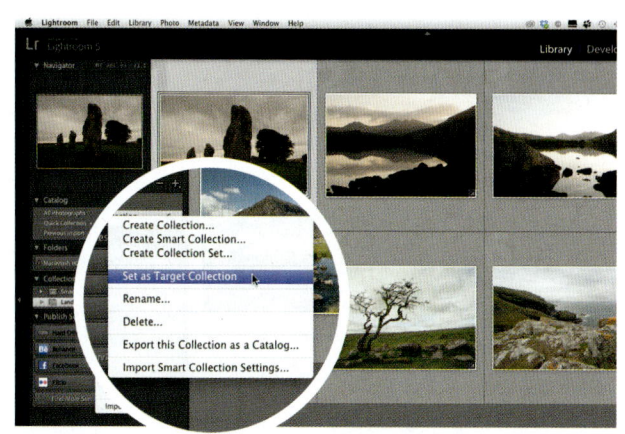

Teach yourself Lightroom
IMPORT AND SORT IMAGES

3 Set as Target Collection
The Quick Collection will empty once you've added the photos to their own Collection. Click on the 'Landscape Collection' to see its contents. Righ-click the Collection label and choose Set As Target Collection. Click the All Photographs option to see all the photos again.

4 Use the Painter tool
Now when you select a thumbnail and press B, the photo will be added to the new Target Collection instead of the default Quick Collection. You can also click the Painter icon and set the Paint drop-down menu to Target Collection. Click the Painter tool on any thumbnail to add it.

5 Create new Collections
To create new Collections, click the + icon to the right of the Collections panel. You can set any new Collections up to be Target Collections. You can also create your own Smart Collections to find files that match specific criteria such as a high ISO setting, for example.

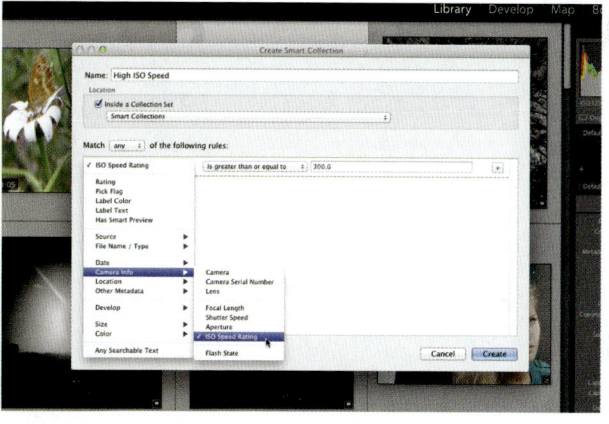

6 Create a Smart Collection
Click + and choose Create Smart Collection. Name it 'High ISO Speed'. Tick Inside Collection Set and choose Smart Collections. Set Match to Any. Change the Rating menu to Camera Info and choose ISO Speed Rating. Choose the Is In Range option and put in a suitable range.

7 See your custom Collection
The 'High ISO' Smart Collection lists all of the photos that match the rules you set. Click the 'High ISO Speed' label and the photos will appear in the grid view. Click on a photo and look in the Metadata panel. This photo will have a high ISO, so it meets the rules you set.

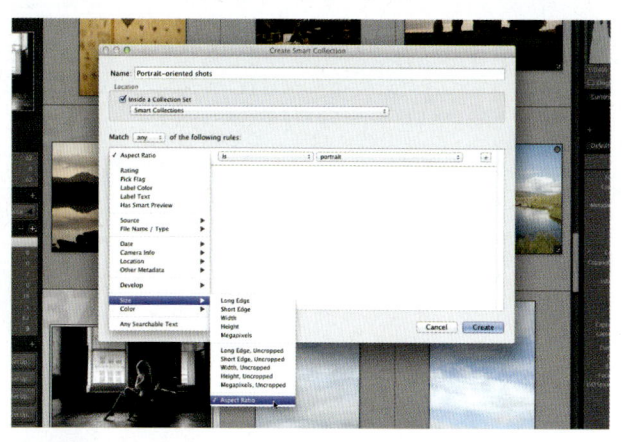

8 Select photos via their shape
Smart Collections can also organise images by their shape. To display only portrait-oriented photos in the Library module, create a new Smart Collection. Scroll down to Size and choose Aspect Ratio. First choose Is from the second menu and Portrait from the third. Click Create.

Teach yourself Lightroom

THE LIBRARY MODULE

Teach yourself **Lightroom**

THE LIBRARY MODULE

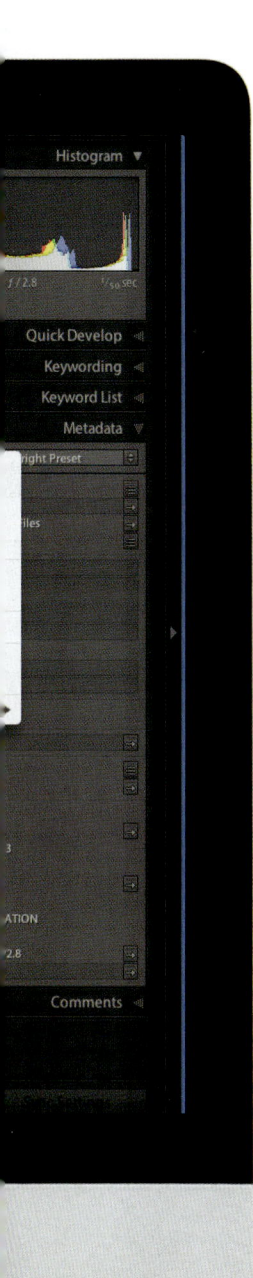

The Library module

Locate, rate, give images keywords and apply quick photo fixes with Lightroom's Library module

20 Introducing the Library module
Discover the key features of the Library module, and customise it to create a cleaner-looking workspace

22 Sort and rate your images in Lightroom
Use the tools in the Library module to start organising your photos to make them easier to manage

24 Find images fast with keywords
Identify particular pictures by adding keywords to the metadata, then search for specific images using keyword filters

26 Tag your images with location data
Use the GPS data embedded in your images to pinpoint the shooting location of a photo in the Map module

Teach yourself Lightroom
THE LIBRARY MODULE

GET THE FILES HERE: http://bit.ly/tylr2016

Introducing the Library module

Discover the key features of the Library module, and customise it to create a cleaner-looking workspace

In the previous chapter we demonstrated how to import images into Lightroom. We also touched on ways to organise your assets by placing them in Collections. In this chapter we'll delve deeper into the Library module and demonstrate how it can help you to manage (and edit) your images. You may have thousands of photos to deal with, and they may be scattered across a variety of folders and external hard drives. Without Lightroom's Library module, you'd have to rummage around in those separate folders looking for specific photos. Lightroom gathers assets together and places them in its Catalog, so that they are all under one roof. You can then see your images more conveniently in the Library module.

On this spread we introduce the key panels and tools in the Library module, so you can familiarise yourself with its layout. We also explain how to change the default layout so you can hide panels until you need them. We'll then move on to show you how to organise your images using keywords and star ratings, and tag them with the coordinates of the location in which they were captured.

At a glance The Library module

Get to know the key features of Lightroom's organising workspace

4 HISTOGRAM

When you click a photo's thumbnail (or double click to view it in the Loupe view), you can see the image's spread of tones in the histogram window. This undulating graph enables you to analyse and fix exposure-related problems, as we'll demonstrate in chapter three.

5 KEYWORDING PANEL

Here you can assign keywords to a photo to help you find it more easily in the future. Like all panels, it can be toggled open or collapsed by clicking the little triangle icon. If you right click a panel you can clear unwanted panels from the pop-up menu.

6 METADATA

Here you can discover lots of information about how a photo was captured, including camera settings such as the ISO and aperture values. You can use Lightroom to add extra information to a photo's metadata, such as keywords and GPS coordinates.

1 FOLDERS PANEL

This panel shows you all the folders linked to Lightroom's Catalog. Right click a folder and choose Synchronize Folder to update it.

2 IMAGE DISPLAY

Photographs and video clips that have been imported into the Library module appear in the image display area as thumbnails.

3 TOOLBAR

The icons in the Toolbar enable you to add ratings, flag photos for rejection and display them by a range of criteria such as Capture Time.

Teach yourself **Lightroom**
THE LIBRARY MODULE

At a glance The Library workspace
Create a cleaner workspace and compare photos in the Survey view

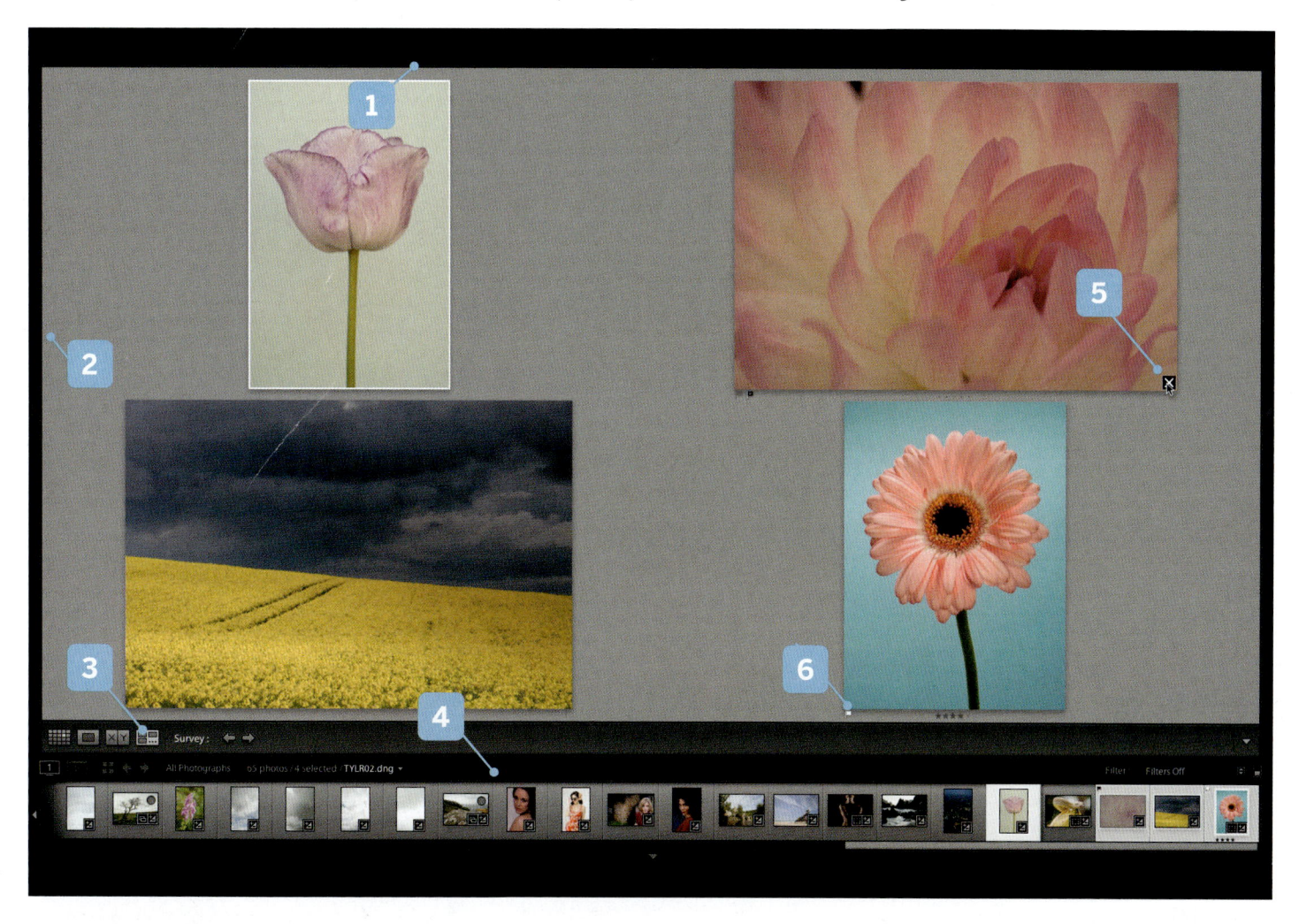

1 HIDE OR SHOW

If you click the little triangle icons by the Library module's side or top panels, you can make them vanish and create more space to work in. To summon a panel, simply slide the mouse to the appropriate edge of the screen.

2 FULL SCREEN

Press Shift+F to cycle through different screen options. Here we've used this shortcut to hide the top menu bar. Press F to make the currently selected thumbnail fill the screen. We've also collapsed the Module picker to make more space.

3 SURVEY VIEW

The Survey view enables you to compare a selection of photos as larger thumbnails. Shift click to select a range of photos to survey. If you only want to compare two photos, click the Compare view icon immediately to the left of the Survey icon.

4 FILMSTRIP

When using the Survey or Compare views, it's handy to be able to access all the Library module's thumbnails too. To do so, click the triangle at the bottom of the workspace to summon the Filmstrip. This offers an alternative to the Grid view.

5 SELECT

Shift click in the Filmstrip to select a range of thumbnails to compare in the Survey view. The first thumbnail selected is the active photo. It will display a white border. To remove a photo from the Survey view, click the X icon.

6 SORT

You can begin to sort your photos by clicking icons or adding ratings. Here we've ticked the Keep flag icon for this photo. We've also added four stars by clicking the icons below the image. There's much more on sorting on the following pages.

21

Teach yourself Lightroom
THE LIBRARY MODULE

GET THE FILES HERE: http://bit.ly/tylr2016

Sort and rate your images

Use the tools in the Library module to start organising your photos to make them easier to manage

If you've imported thousands of photos into Lightroom's Library module, you may be daunted by the challenge of finding specific files when you need them. Fortunately, Lightroom is packed full of tools and commands that are designed to help you separate the wheat from the chaff. In chapter one we demonstrated how to add selected photos to Collections, so you could group images together in a similar way to putting prints into a photo album. We also demonstrated how to use the Smart Collections feature to collect and display files according to specific metadata criteria, such as whether they had a portrait or landscape aspect ratio, for example. Lightroom displays all your thumbnails according to the date they were captured, so you can scroll back through time and rely on your memory of when particular images were taken. As this chapter progresses, we'll show you alternative and much more efficient ways to organise and search for files.

We'll start by examining quick and easy ways to sort your photos using ratings and flags, so you can begin the quality-control process by highlighting your favourites, while marking others for rejection. We'll also show you how to filter the photos in the Grid view to display images that match specific criteria contained in each photo's metadata. We'll then move on later in this chapter to explore more advanced ways of organising photos, using tools such as keywords and geo-tagging.

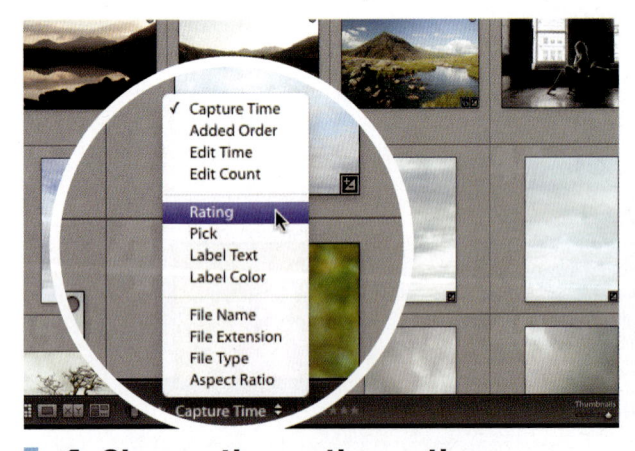

1 Change the sorting options
Press G to display the Grid view. Photos are listed by when they were captured, with the newer thumbnails at the top. From the Toolbar at the bottom you can use the Sort pop-up menu to change the way photos are displayed, such as making the highest-rated photos appear at the top.

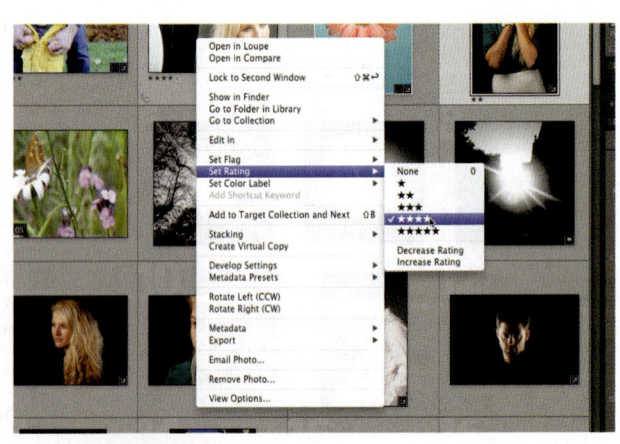

2 Rate your photos
To rate a photo, click its thumbnail, then click below the thumbnail to assign between zero and five stars. Alternatively, you can tap one of the number keys or click a star rating icon in the Toolbar. You can also right-click a thumbnail and use the Set Rating command.

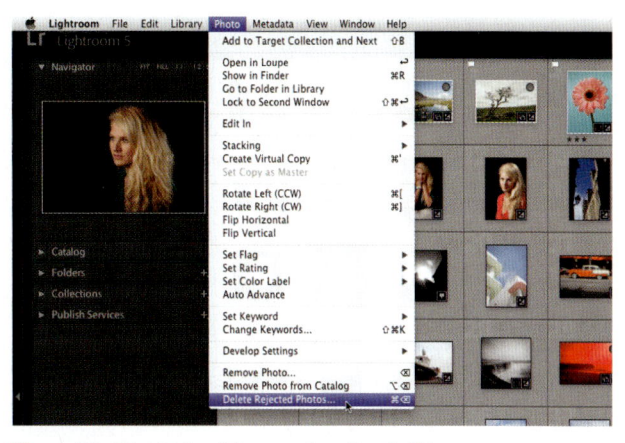

Teach yourself Lightroom
THE LIBRARY MODULE

3 Check the quality
Double-click a thumbnail (or press E) to take the photo into Loupe view. You can then click the image to zoom into a 1:1 view that will show you the image at 100%. If you want to keep the photo, tick the Set as Pick flag icon in the Toolbar. Otherwise, click the Set as Rejected flag.

4 Delete the rejected files
You can use the Toolbar's Sort menu to display the picked pictures at the top of the Grid view. Photos that have been flagged as Rejects will appear as greyed out thumbnails at the bottom of the Grid view. Choose Photo>Delete Rejected Photos to remove them.

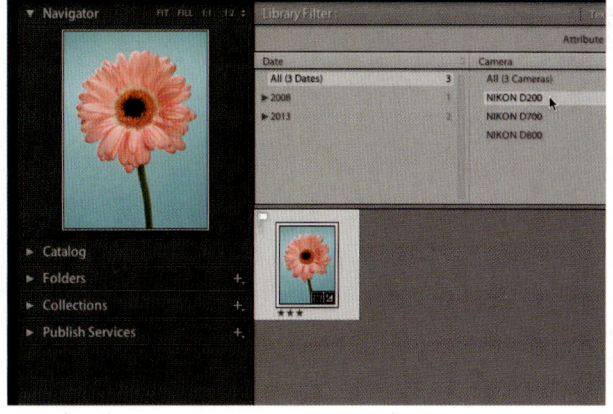

5 Remove or delete?
When you choose to delete a photo, you have two options. If you click Remove, it will stay in its original folder on your PC, but it will vanish from Lightroom's Catalog. To permanently delete it from its original folder, click the Delete from Disk option.

6 Sort using filters
Go to View>Show Filter Bar. Tick Attribute. Click the Pick icon to display photos that have been flagged. You can also choose to display photos that have been rated greater than or equal to a specific star rating. You can refine the filter results by clicking Metadata and choosing an attribute.

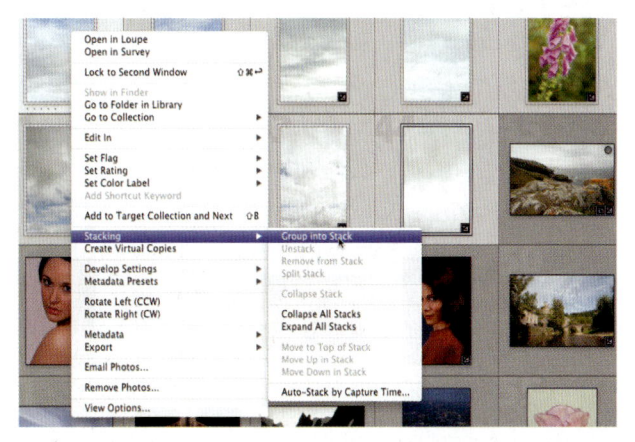

7 Colour-code your pictures
The Filter Bar also displays photos according to colour labels. Click the Custom Filter menu and choose Filters Off. This will display all the photos. To assign a coloured label to a photo, simply right-click its thumbnail, choose Set Color Label, and choose a colour from the list.

8 Stack images to save screen space
You may have a collection of similar photos cluttering the workspace. To save space, Shift-click each similar thumbnail, then right-click and choose Stacking>Group into Stack. You can then click the right of a collapsed stack's thumbnail to expand it, or left to collapse it.

Teach yourself **Lightroom**
THE LIBRARY MODULE

GET THE FILES HERE: http://bit.ly/tylr2016

Find images with keywords

Identify particular pictures by adding keywords to the metadata, then search for specific images using keyword filters

We just looked at ways to begin sorting your collection of photos by adding ratings and flags, and using their existing metadata to filter photos according to certain criteria, such as aspect ratio or camera model. Later in this chapter we'll learn how to locate photos according to where they were captured, thanks to the GPS coordinates stored in the metadata.

It's handy to be able to search for files according to coloured labels or shooting locations, but these tools will only go so far. To find a photo according to its actual subject matter, we could try to remember when it was taken and then scroll down through the Grid view in an attempt to identify its thumbnail. Alternatively, we could add keywords to our photos and then use Lightroom's Filter bar to find the particular pictures in an instant. Keywords are a series of words that describe the content of a photo. They are stored in the photo's metadata, so that when the file is shared or uploaded, the keywords remain attached to it. This can be especially useful if you want to sell your work as stock photography, because potential buyers can use keywords to find your photo via search engines.

In this walkthrough we'll show you how to create keywords from scratch, and then use the handy Keyword Set panel to assign common keywords to photos with just a few clicks. It may seem unnecessary if you don't have many photos to begin with, but the more effort you put into adding keywords to your photos, the easier they will be to find in the future. Let 'Keywords are King' be your image-organising mantra!

Teach yourself Lightroom
THE LIBRARY MODULE

1. Assign keywords on import
You can assign keywords to a batch of photos when you import them into Lightroom's Catalog. Go to the Apply During Import panel and type into the Keywords field. Separate each keyword with a comma. Lightroom will remember previous keywords and suggest them as you type.

2. Keyword Sets
In the Library module, double-click to select a landscape photo. To see a photograph's Info label, press Cmd/Ctrl+I. Toggle open the Keywording panel. Click the Keyword Set drop-down menu and choose an appropriate set such as Outdoor Photography.

3. Manually assign keywords
Click a preset keyword such as 'Landscape' to assign it to the selected photo. The assigned keyword appears in the Keyword tags panel. Click 'Spring'. That word will also appear in the Keyword tags panel, separated by a comma. You can then manually type more specific words such as ocean.

4. Use keyword suggestions
Press G to go to the Grid view and double click a similar photo. You'll see more keyword suggestions (based on previously added keywords) in the Keyword panel. Click appropriate keyword suggestions to add them to the selected photo's metadata.

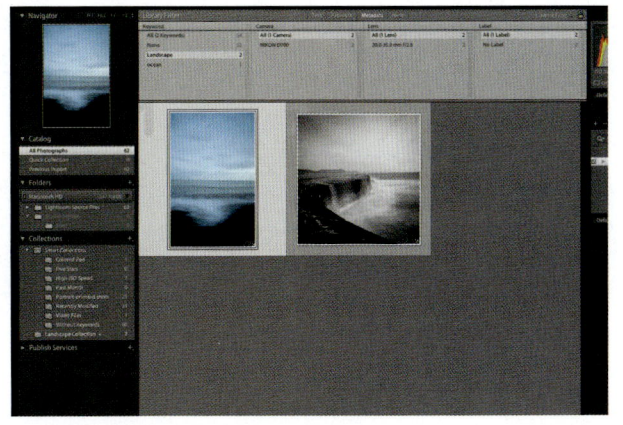

5. Find keyworded photos
Toggle open the Keyword List panel. Here you'll see a list of all the keywords that have been added to photos in Lightroom's Catalog. You can also see how many photos contain a particular keyword. Place the cursor on a keyword to highlight the appropriate thumbnails with a white border.

6. Filter by keywords
If you click the white arrow on the right of a keyword (such as 'Landscape') it will summon the Library Filter. The keyword 'Landscape' will be highlighted in the Library Filter's Keyword panel and only photos containing that keyword will be visible in the Grid view.

Teach yourself Lightroom
THE LIBRARY MODULE

GET THE FILES HERE: http://bit.ly/tylr2016

Tag your images with location data

Use and add GPS data embedded in your images to pinpoint the shooting location of a photo in the Map module

After travelling abroad it used to be traditional practice to print your holiday snaps and store them in an album to flick through. Now, you can gather your holiday-sourced photos together in a Lightroom Collection (see chapter one). Alternatively, you can use the Map module to present your photos according to where they were captured, which provides a fun and informative way to find particular files.

When you take a photo on your smart phone, the software records the location of the photo using GPS (the satellite-based global positioning system). These coordinates are stored in the image file's metadata, so when you import it into Lightroom the software can read the map coordinates and display the photo's location in the Map module. Here you can see a map, see little flags that indicate a geo-tagged image, and click to see any photos that were taken there.

To demonstrate how the Map panel works, we've provided you with some iPhone-sourced photos that contain geo-tagged location coordinates. Make sure that you import the supplied iPhone-sourced images into the Library module so you can follow our walkthrough. If your camera doesn't automatically geo-tag your photos then we'll demonstrate how you can manually add location data to a photo by dragging and dropping it onto the appropriate part of a map.

1 Find your photos
Set the Custom filter drop-down menu in the Library Filter to Camera Info. On the Camera tab, click on your desired camera to display photos taken. Photos with GPS data already built-in will display a GPS icon. Click on a photo to see its GPS information in the Metadata panel.

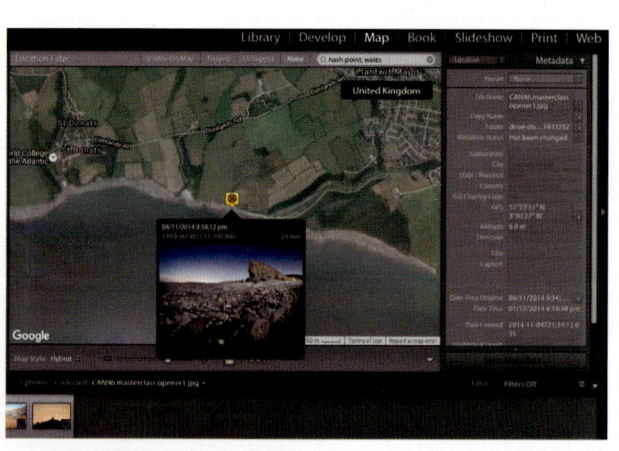

2 Tag a location
In the Map module, type a place name into the search bar. If the place name returns no results then find it on the map and zoom in with the + button at the bottom of the map. Drag your image's thumbnail onto the map to assign a location. GPS coordinates will appear in the photo's metadata.

Teach yourself **Lightroom**
THE LIBRARY MODULE

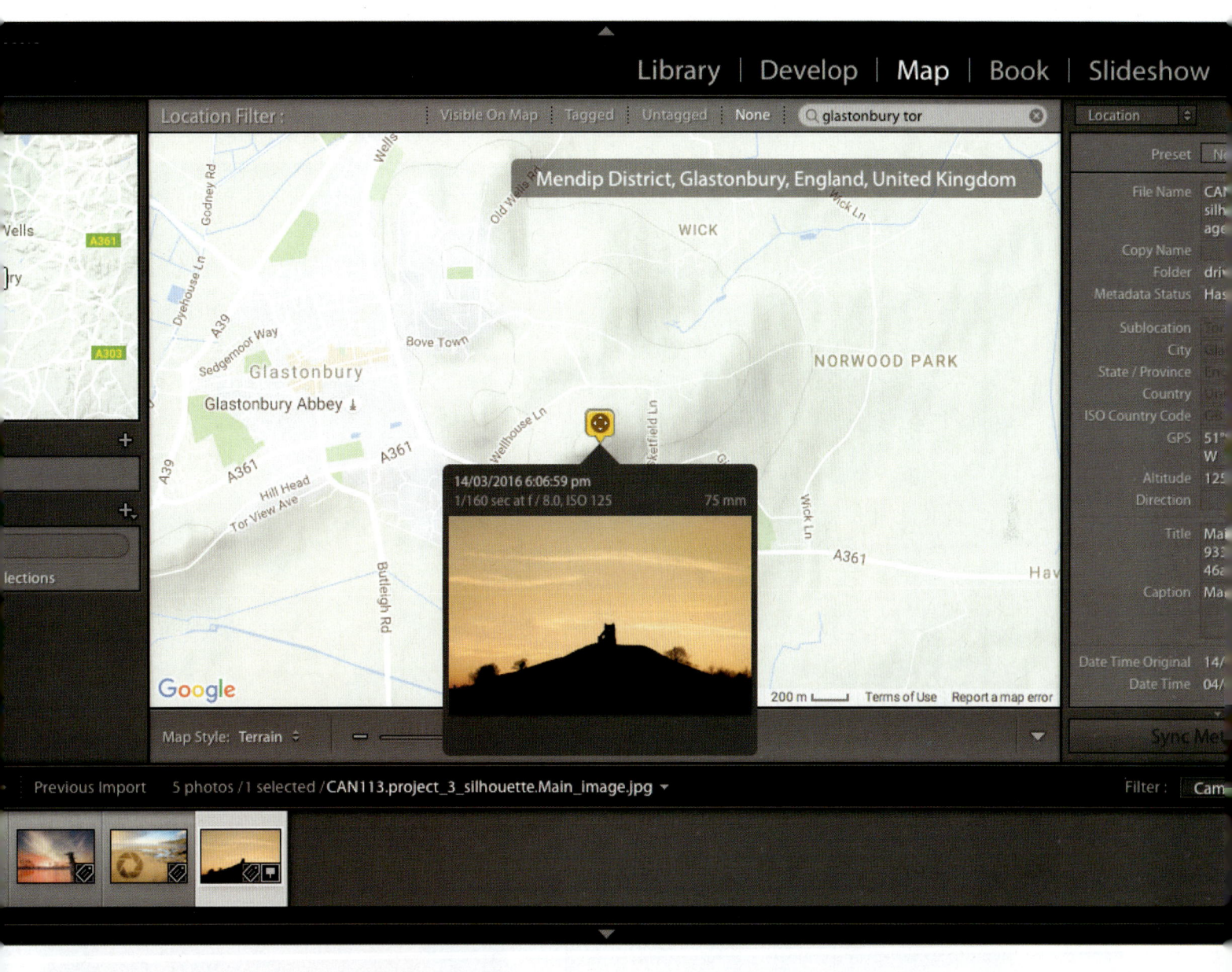

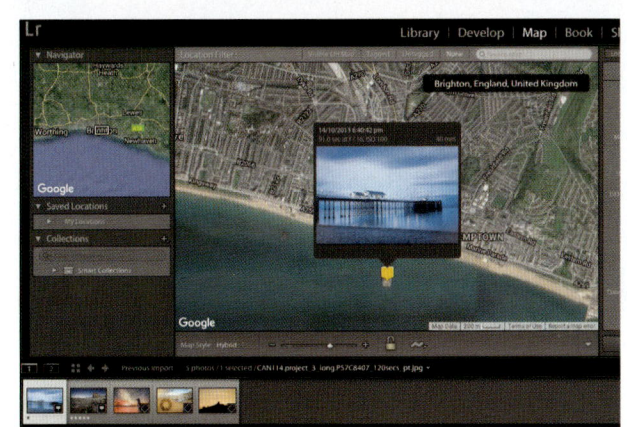

3 View by location
Click the arrow to the right of the GPS coordinates and you'll jump to that location in the Map module. The photo's position will be indicated by a yellow flag. Place the cursor over a flag to see the tagged photo. Use the arrow keys to cycle through multiple photos.

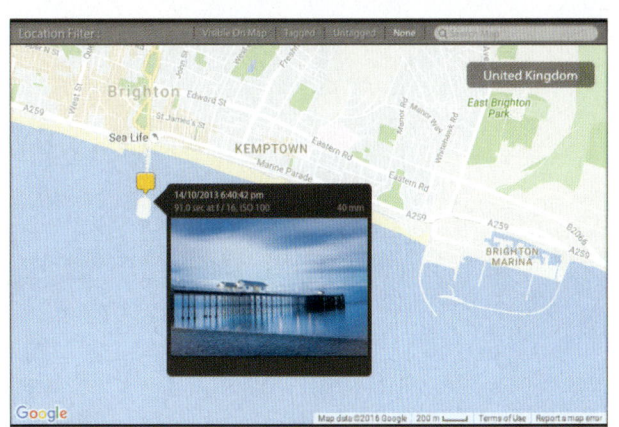

4 Filter the flags
Use the options in the Location Filter to highlight particular photos (such as those tagged with GPS coordinates) in the film strip. As you move over a thumbnail in the film strip, the associated flag will jump up and down on the map. Use Map Style to change the look of the map.

Teach yourself **Lightroom**
THE DEVELOP MODULE

Teach yourself **Lightroom**
THE DEVELOP MODULE

The Develop module

Discover how to make basic adjustments to images quickly and easily using the Develop module

30 Introducing the Develop module
You'll find Lightroom's key image-editing tools and photo-fixing panels in the Develop module

32 Discover how to use the histogram
Use Lightroom's histogram window to quickly and easily diagnose and fix tonal problems in your digital photos

34 Use Smart Previews to edit remote images
Edit images in the Lightroom Catalog that aren't currently accessible on your computer using Smart Previews

38 Make basic colour adjustments
Use Lightroom's colour-correcting tools to boost weak colours without over-saturating strong ones

40 Crop and straighten your photographs
Use Lightroom's Crop Overlay tool to straighten tilted horizons and improve the composition of your photos

Teach yourself Lightroom
THE DEVELOP MODULE

GET THE FILES HERE: http://bit.ly/tylr2016

Introducing the Develop module

You'll find Lightroom's key image-editing tools and photo-fixing panels in the Develop module

At the end of the previous chapter on the Library module, we touched on ways to improve photos using the Quick Develop panel. This aptly named panel enables you to quickly tweak some of the most common image-related issues. Once you've imported and organised your collection of digital negatives into Collections and added keywords, you'll be ready to deal with more complex image-processing challenges — and this is where the sophisticated Develop module comes in.

Adobe invented the .dng (digital negative) format to make it easier to store and access unprocessed raw files. The word 'negative' harks back to the traditional days of analogue film. Before you could see a film negative's true colours and tones, it needed to be developed in a darkroom, which is why the Develop module's name also evokes recollections of traditional darkroom techniques. Here you can process a picture's colours, tones and composition to perfection. You can also make selective adjustments to your raw files, so that even high-contrast scenes can feature detail in the shadows, mid-tones and highlights.

At a glance The Develop module
Get to know the key features of Lightroom's digital darkroom

1 NAVIGATOR
This panel enables you to get a closer look at your picture. Click on a size at the top to zoom in by a specific magnification.

2 HISTOGRAM
This panel displays the spread of tones in a photo. Here we see some strong shadows on the left, and an absence of strong bright highlights.

3 TOOLBAR
These darkroom-style tools enable you to improve your photos with a few clicks. You can crop images, remove sensor spots, or make tonal edits.

4 BASIC
The Basic panel enables you to tackle the most common picture problems, such as under-exposed shadows or colour casts produced by a camera's incorrect white balance setting. We'll look at this important panel in more detail on the opposite page.

5 SOLO MODE
When multiple panels are open, the workspace can become cluttered. Right-click a panel to summon this pop-up menu. Hide unwanted panels by un-ticking their labels. Solo Mode enables you to click a panel to open it. It will then close all the other panels that are open.

6 PRESETS
These presets enable you to make many adjustments with a single click. They change a multitude of settings in Develop module panels to produce a range of creative or photo-fixing edits. Check out chapter seven for more on this powerful panel.

Teach yourself **Lightroom**
THE DEVELOP MODULE

At a glance The Basic panel
Familiarise yourself with the Basic panel's key features

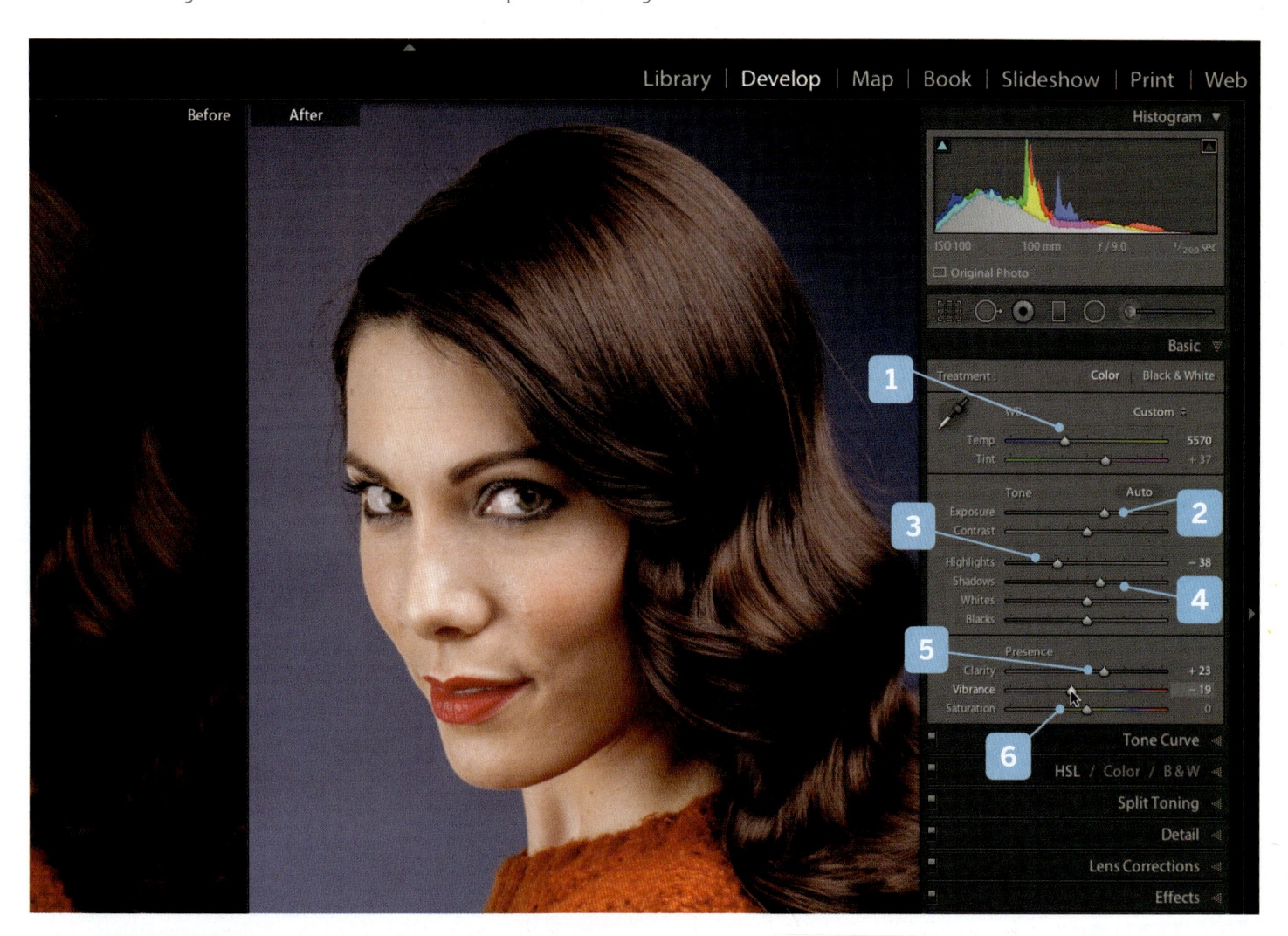

1 WHITE BALANCE
This section enables you to try out different white balance presets (Such as Flash or Daylight), or manually change the colour temperature. Here we've dragged the Temperature slider down to 5570 to cool down the starting image's rather warm colour cast.

2 EXPOSURE
Our starting image is under-exposed and lacks detail. By dragging Exposure to +1.20 we can open up the aperture to reveal more detail. This adjustment brightens the photo's tones and causes the histogram graph to shift to the right.

3 HIGHLIGHTS
The global adjustment made by the Exposure slider has caused the brightest pixels to over-expose. We can selectively claw back detail in the brightest highlights by dragging this slider left to -38. This adjustment doesn't alter the darker tones.

4 SHADOWS
This slider enables you to lighten under-exposed shadows and reveal more detail. It doesn't affect the brighter highlights and white pixels towards the right of the histogram window. It also leaves the photo's darkest black pixels untouched at the far left of the graph.

5 CLARITY
This clever slider increases midtone contrast, which is a great way to make fine details such as our subject's hair stand out. It does a similar job to the sharpening tools in the Detail panel. We'll look at Clarity in more detail in the next chapter.

6 VIBRANCE
This slider enables you to selectively adjust saturation. It can boost the strength of weaker colours without over-saturating stronger ones. Here we've reduced the strength of the vivid orange jumper without desaturating the skin tones too much.

Teach yourself Lightroom
THE DEVELOP MODULE

GET THE FILES HERE: http://bit.ly/tylr2016

The histogram

Use Lightroom's histogram window to diagnose and fix tonal problems in your digital photos

The histogram window is one of the most useful tools in the Develop module. The histogram graph performs a similar function to the medical chart clipped to a hospital patient's bed. Just as a doctor or nurse can use a chart to help diagnose the health of a patient, we can look at the shape of a photo's histogram graph to see if the image is under- or over-exposed.

We can also use Lightroom's histogram window to help us create an image that has a healthier spread of tones that displays more detail in its shadows, midtones and highlights. In the walkthrough below we'll show you how to analyse a histogram so that you can understand how parts of the undulating graph equate to a photo's tonal characteristics.

Our TYLR04.dng starting image looks over-exposed to the naked eye. We'll show you how to use the histogram to create a correctly exposed photo. To get started, first import TYLR04.dng from the download files into Lightroom and open it in the Develop module...

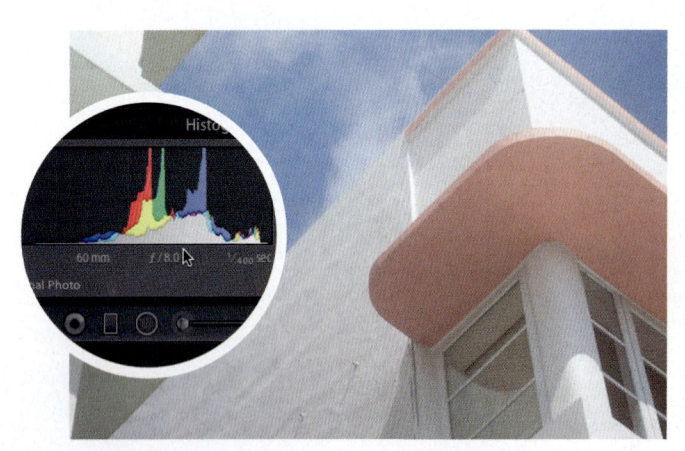

1 Analyse the histogram
The graph is bunched up towards the middle and right of the histogram window. This indicates that the image consists mostly of midtones (in the middle) and highlights (on the right). The fact that there's no graph data on the far left demonstrates the photo's lack of shadows.

2 Darken the midtones
A well-exposed photo should have a graph that spreads from the blacks on the far left to the whites on the far right. Place the cursor over the peaks in the middle. The Exposure slider will be highlighted. Drag left on the graph to reduce Exposure to -0.60. This darkens the midtones.

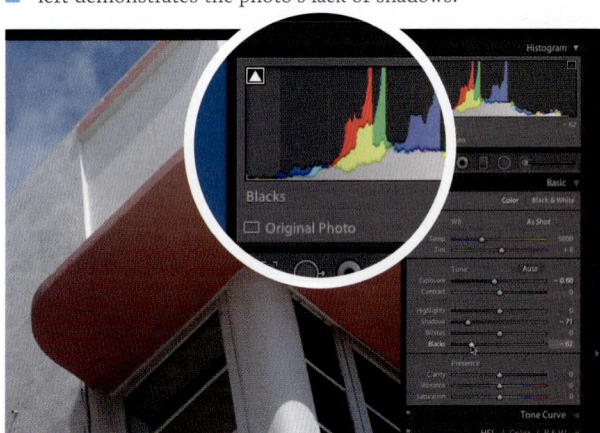

3 Show the clipping warnings
Drop Shadows to -71. A photo should have some black pixels, so drag Blacks left to -62. Click the Show Shadow Clipping icon to see the blackest shadows as patches of blue. These clipped areas will print as pure black, but there's no crucial detail being lost in these small sections.

4 Improve the contrast
For a strong contrast we need some white highlights. Turn on Show Highlight Clipping. Drag Whites to +50. A few red patches will appear when you begin to blow up detail in the brightest highlights, but again, there's no important detail to lose in these areas.

Teach yourself **Lightroom**
THE DEVELOP MODULE

AFTER

BEFORE

33

Teach yourself Lightroom

THE DEVELOP MODULE

GET THE FILES HERE: http://bit.ly/tylr2016

Use Smart Previews

Edit images in the Catalog that aren't currently accessible by Lightroom using Smart Previews

As we revealed in chapter one, Lightroom can gather photographs from a variety of locations (such as external hard drives) into its Catalog. It then creates and stores standard-resolution previews of each imported photo on your PC's hard drive, while storing the original high-resolution images on your external hard drives. This enables you to save valuable disc space on your laptop or PC. However, if you plan to use a laptop to edit your photos while away from the office, then you won't want to have to lug around a collection of hard drives in order to access and edit your images.

By creating a Smart Preview of a photo, you can edit the image while on the move, without requiring access to the external hard drive. After you disconnect the external hard drive containing the original raw file, you can still apply all of the photo-fixing tools in the Develop module to the Smart Preview version of the photo. When you reconnect your laptop with the external hard drive, Lightroom will automatically look at all the edits you made to the Smart Preview, and then apply them to the high-resolution original.

In the following walkthrough we'll work with an overexposed starting image that's been imported and stored on an external hard drive. After downloading our TYLR09.dng starting image, copy it onto an external drive too, so you can discover how Smart Previews work their magic to give you the freedom to edit any photo at any time and in any place.

Teach yourself **Lightroom**

THE DEVELOP MODULE

35

Teach yourself Lightroom
THE DEVELOP MODULE

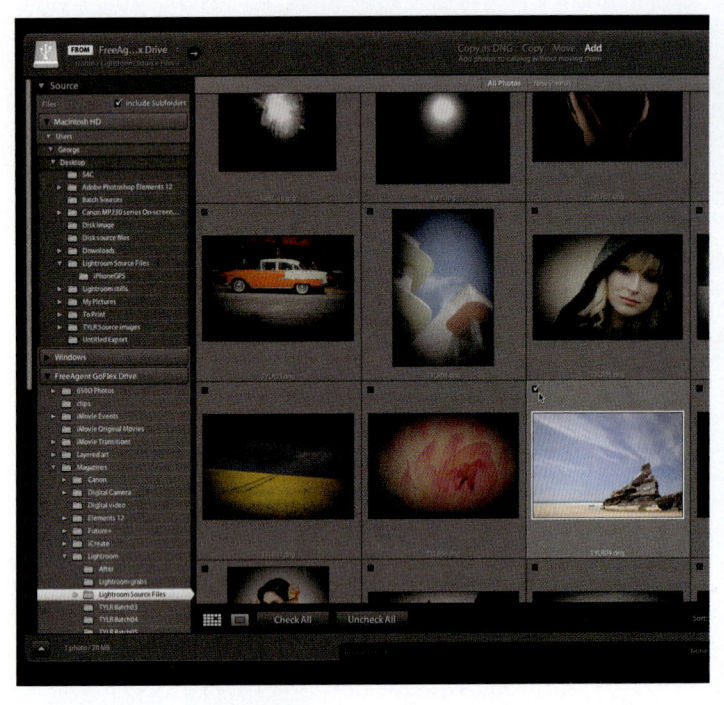

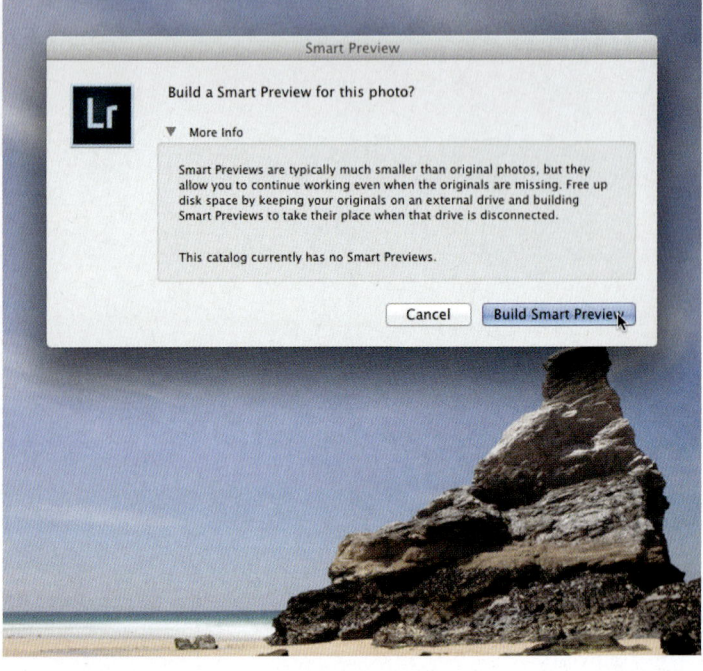

1. Import a photo
Go to File>Import Photos and Video. Browse to the external drive in the Source section of the import window and select the TYLR09.dng thumbnail. Click the Add button so that Lightroom's Catalog will link to the photo without moving it onto your PC. Leave the Build Smart Previews box un-ticked at this stage, because you can always choose this option for specific photographs at a later date. Click Import.

2. Standard versus Smart
By default Lightroom will create a standard preview of any imported photo so that you can see it in the Library module. The original high-resolution raw file will remain on the external drive. However, if the external hard drive isn't connected then you won't be able to edit a standard preview version of an image. Go to the histogram window and click the Original Photo box. A Smart Preview dialog will appear.

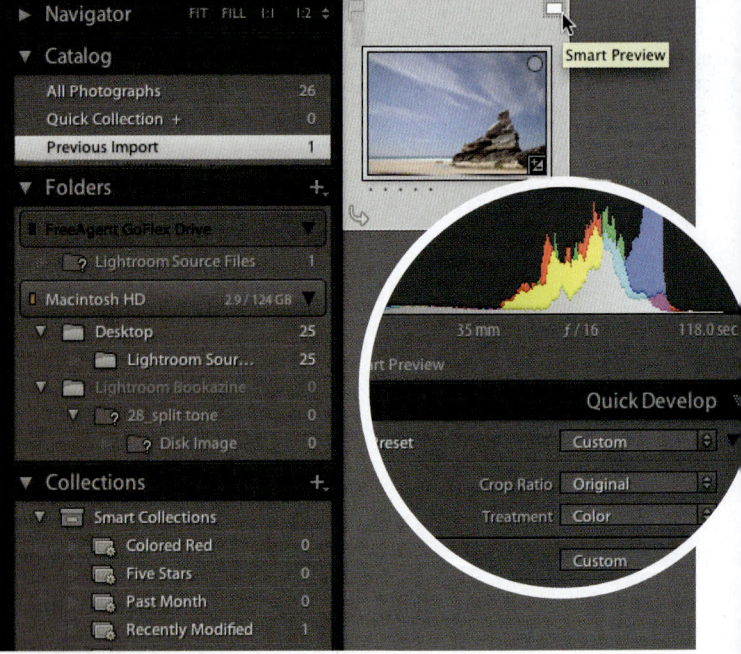

3. Build the Smart Preview
Click the Build Smart Preview button. After a short while a new dialog will appear indicating that 'One Smart Preview was built'. Click OK to dismiss the window. Look below the histogram window. A new label will indicate that you now have the original file plus a Smart Preview version. As an experiment, eject the external hard drive by disconnecting it from your PC. Now only the Smart Preview label will be visible below the histogram.

4. Dismiss the warning
If you click the Smart Preview label below the histogram graph, a new dialog will appear indicating that the photo is missing or offline, but it can still be edited using its Smart Preview. Click OK to dismiss the warning. Any other standard preview photos in the Library that were on the external drive will feature an '!' (photo missing) symbol. If the external drive is still disconnected, then our TYLR09.dng thumbnail will have a Smart Preview icon on it.

Teach yourself **Lightroom**

THE DEVELOP MODULE

5 Analyse the histogram
Take the photo into the Develop module. You can see from the spread of tones on the histogram graph that there are lots of strong midtones in the middle and a few highlights towards the right, but there's very little shadow information on the left. This indicates that the photo is slightly over-exposed. We need to increase the strength and spread of the histogram graph towards the left to create stronger shadows.

6 Improve the contrast
To improve the photo's contrast and create a wider spread of tones, drag the Exposure slider left to -0.20 to gently darken the photo overall. Drag the Blacks slider left to -23. This makes the histogram stretch to the far left, so we now have some dark black pixels present in the picture. For some contrasting white highlights, drag Whites right to +39. This adds a hint of clipping to some of the foam, but no crucial detail is lost.

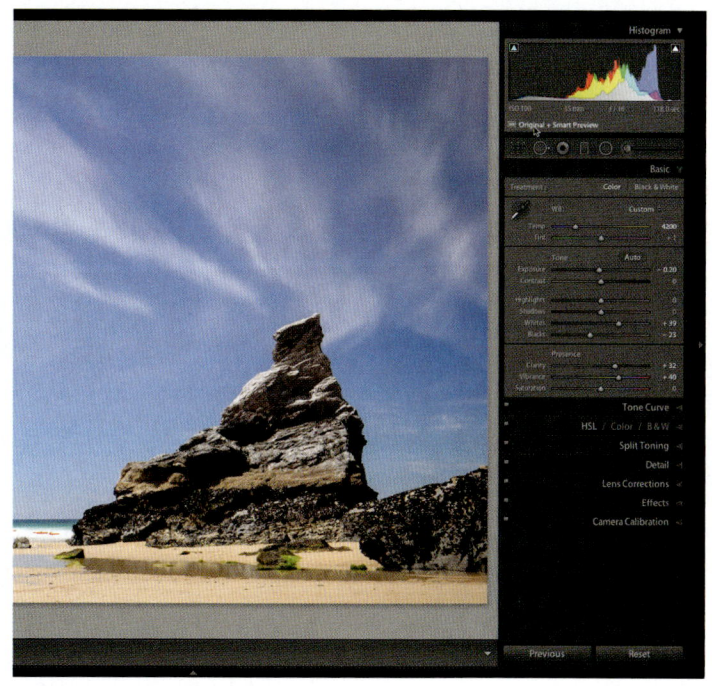

7 Selectively boost the colours
To tease out some of the texture in the rocky midtones, drag Clarity right to +31. This slightly increases the spread of the foam's clipped highlights, but detail in these areas is negligible. By boosting Vibrance to +40 you can selectively increase the saturation of natural-looking colours such as the blue sky and green seaweed. Warmer colours tend not to be altered as much because this is designed to stop any skin tones from becoming too orange.

8 Reconnect the drive
Once you've finished editing your Smart Preview, reconnect your external hard drive. Any 'photo missing' warning icons by other thumbnails will disappear. If you look below our TYLR09.dng photo's histogram window you'll see that the 'Original+Smart Preview' label has reappeared. All the colour and tonal changes that we made to the Smart Preview will be applied to the original raw version of the photo on the external drive.

Teach yourself **Lightroom**
THE DEVELOP MODULE

GET THE FILES HERE: http://bit.ly/tylr2016

BEFORE

AFTER

Make basic colour adjustments

Use Lightroom's colour-correcting tools to boost weak colours without over-saturating strong ones

If you plan to share your edited photographs online then you don't need to worry as much about over-saturating the colours. Computer screens can reproduce millions of colours, but if you're planning to print your photos then you need to be more careful about avoiding over-saturated colours at the editing stage, because printers can't reproduce as many colours as you can see on a monitor. If you print over-saturated colours, they may look garish and lack detail in the print version. We'll look into ways to get more faithful-looking prints in chapter nine, but it's worth being aware of the concept of printable and unprintable colours at an early stage.

Lightroom is also designed to help you counteract colour casts. These warm orange or cold blue tints can occur when your camera's automatic white balance setting fails to get whites looking white in particular lighting scenarios. We'll look at colour correction in more detail in the following chapter, but it will be helpful to get to grips with Lightroom's colour-correcting tools as soon as possible because you'll need to use them often.

Every edit that you perform in Lightroom's Develop module is stored in the History panel. This enables you to perform multiple undoes if you need to retrace your steps, or you can jump straight to a particular stage in the editing process by clicking the appropriate history state. You can also record versions of your edited image as easily accessible Snapshots, which gives you the freedom to experiment with different looks.

Teach yourself Lightroom
THE DEVELOP MODULE

1 Adjust the saturation
Open TYLR13.dng in Lightroom's Develop module. The photo looks a little desaturated, apart from the bright orange lichen on the rock. Go to the Presence section of the Basic panel. If you decrease the value of the Saturation slider to -100 then all the colours become greyscale.

2 Adjust the Vibrance
Restore Saturation to 0. Reduce Vibrance to -100. The weaker colours become grey, while the stronger orange lichen retains some colour. This demonstrates how the Vibrance slider selectively adjusts colour saturation. It also has a stronger influence over typical landscape colours.

3 Selectively boost the colours
If you increase Saturation, the lichen becomes too saturated and unprintable. To selectively boost the greens and blues without over-saturating the stronger oranges, drag the Vibrance slider right to a value of +74. This gives the blue ocean and the green moss more impact.

4 Create a Snapshot
Before making further colour adjustments, let's record this version as a Snapshot. Go to the Snapshots panel on the left of the Develop module's workspace. Click on the + icon to create a new Snapshot. It names the Snapshot using date and time. Re-label it 'Vibrance adjustment'. Click Create.

5 Adjust the white balance
The 'Vibrance Adjustment' Snapshot will appear in the Snapshots panel. To enhance the ocean's blues, drag the Temperature slider to 5066. This makes the green sea look more blue. Counteract the green tint by dragging Tint to +15. Create a new Snapshot called 'White Balance'.

6 Compare the versions
You can now click the Snapshot labels to toggle between the two versions of the photo. The Vibrance Adjustment Snapshot has a greener hue than the White Balance version. You can create as many Snapshots as you like to help you decide which edited version you prefer.

Teach yourself **Lightroom**
THE DEVELOP MODULE

GET THE FILES HERE: http://bit.ly/tylr2016

Crop & straighten your photographs

Use Lightroom's Crop Overlay tool to straighten tilted horizons and improve the composition of your photos

While looking at a scene through your camera's viewfinder, you can zoom and pan the lens to compose the photo. You may know what makes a good composition, or you might employ a more intuitive point-and-shoot approach. Often, it's only when looking at photos on the camera's LCD display (or even in Lightroom's Library) that you can spot a photo that stands out as being well-composed.

When taking the picture you may not notice a distracting object at the edge of the frame. For example, when directing a model in the studio you'll be more attentive to her pose than concentrating on any bits of backdrop or lighting kit that might be straying into the photo. And you may not notice problems such as tilted horizons until you've imported a photo into Lightroom. Thanks to the Crop Overlay tool you can remove distracting objects with ease, straighten wonky horizons and even turn a landscape-oriented photo into a portrait-oriented one. Due to the high resolution produced by DSLR cameras you can use the Crop Overlay to make certain features look more prominent in the frame and still produce large prints from a cropped photo. Cropping raw files in Lightroom is also non-destructive, so you can always restore the cropped photo's original composition.

By activating the Crop tool's range of overlays you can recompose your photographs so that their contents adhere to some of the classic rules of composition that have been used by master artists for centuries. We'll take a look at those handy overlays in more detail in our walkthrough.

BEFORE

Teach yourself **Lightroom**
THE DEVELOP MODULE

AFTER

Teach yourself Lightroom
THE DEVELOP MODULE

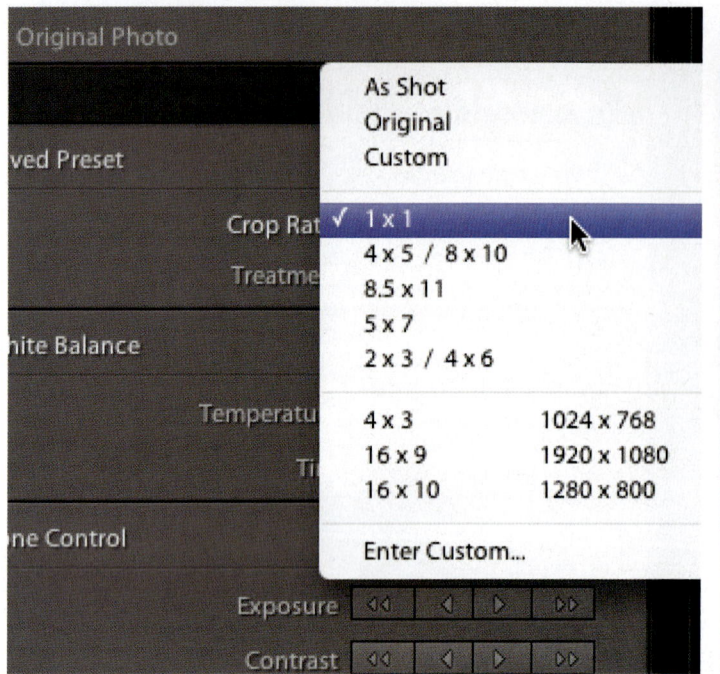

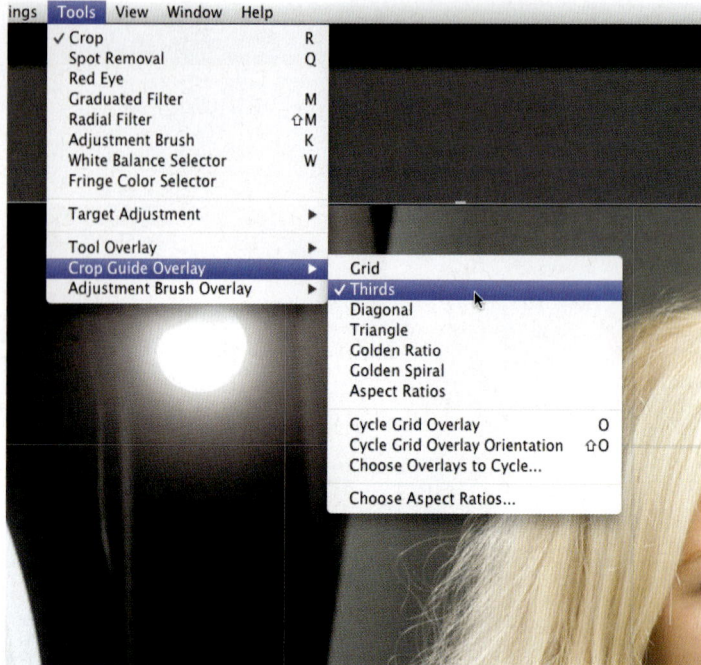

1 Use a preset crop
Import our TYLR33.dng starting image into Lightroom's Library. In Quick Develop you can access and apply preset crops by clicking the Crop Ratio drop-down menu. Some of these presets (such as 5x7) subtly change the photo's shape (aspect ratio). Others (such as 1x1), make a more dramatic change (by making a square image). A square crop doesn't suit this particular image, so revert back to the As shot option.

2 Choose an overlay
For more control over the crop, take the photo into the Develop module. Click the Crop Overlay icon on the left of the tool panel. You can summon different overlays to help you crop subjects. Go to Tools>Crop Guide Overlay and explore the options. Thirds summons a rule-of-thirds grid that helps you to place objects in a balanced composition. Grid is especially handy when looking for unwanted tilted horizons.

3 Create a landscape crop
To keep the cropped photo's shape the same as the original, tick the padlock icon to make sure that it's in a locked position. Drag the top-left corner handle of the Crop Frame tool down to the right to remove the lamp on the left-hand side of the frame. Drag inside the overlay to reposition it so that we can see more of the model's head. This creates a tight landscape-shaped crop that has the same aspect ratio as the original. Click Done.

4 Create a portrait crop
Our photo's subject doesn't suit a landscape-shaped crop, however, so click the Crop Overlay icon to reveal the Crop Frame tool's overlay. Drag the top-left handle up to the right, and the overlay will change its landscape shape to a portrait-oriented one. Drag the corner handles to create a tighter crop. She now looks towards the space on the right of the frame. The unbalanced distractions on the left are hidden. Click Done.

Teach yourself **Lightroom**

THE DEVELOP MODULE

5 Keep the shapes consistent
Because we kept the padlock icon ticked, the cropped photo still has the same aspect ratio (or shape) as other portrait-oriented photos produced by the same camera, even though it started out with a landscape orientation. This means that the photo will look consistent when displayed with un-cropped portrait photos captured during the same shoot, so you can create a series of prints or an online gallery with pictures that all have the same shape.

6 Create a crop Snapshot
On the previous spread we recorded different colour treatments of the same image as handy Snapshots so that we could experiment with different looks, using the Snapshots to compare them. You can store different cropped versions as Snapshots too. Go to the Snapshots panel and click the + icon. Label the new Snapshot 'Portrait version' and click Create. A new 'Portrait version' Snapshot label will appear in the Snapshots panel.

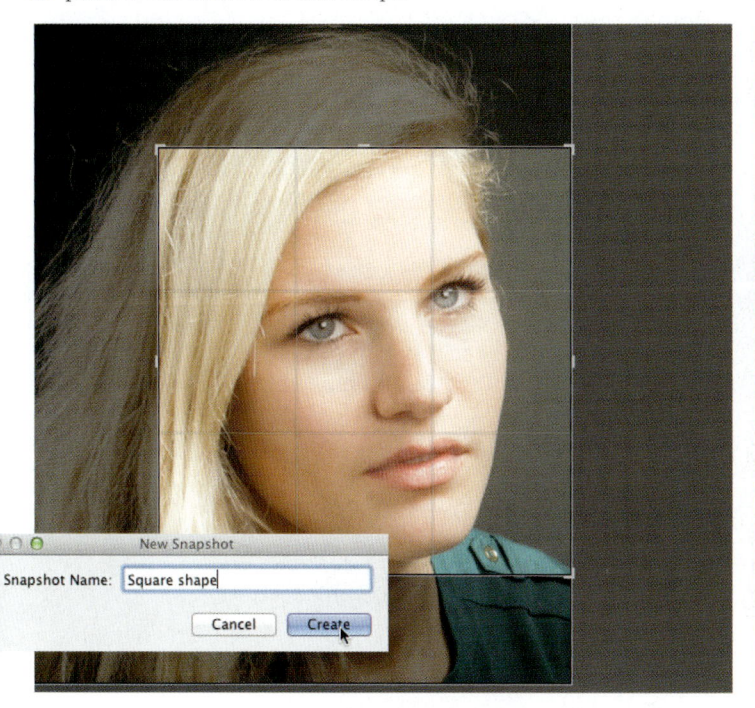

7 Create different compositions
Click the Crop Overlay icon. Click the As shot drop-down menu next to the padlock and choose a preset size such as 1x1. This gives you the square overlay we saw in step one, but this time you can drag inside the overlay to position the face in the square shape. Click Done. Create a new Snapshot called 'Square shape'. You can then toggle between the cropped versions to see which composition you prefer.

8 Straighten an image
Import TYLR24.dng and take it into the Develop module. Click the Crop Overlay icon. Choose Tools>Crop Guide Overlay>Grid. You can easily see that the horizon is tilting down to the right. Click the Straighten tool icon. Draw a line that follows the tilted horizon. The tool will automatically rotate the overlay to counteract the tilt. You can fine-tune it by dragging the Angle slider, if necessary. Click Done to make the crop.

43

Teach yourself **Lightroom**
PHOTO-FIXING TOOLS

Teach yourself **Lightroom**

PHOTO-FIXING TOOLS

Photo-fixing tools

Improve your images quickly and easily using the sophisticated tools in the Develop module

46 **Remove distortion caused by your lens**
Set up and apply lens profiles in Lightroom to automatically counteract lens-induced distortion

48 **Get a better perspective in your images**
Straighten converging verticals caused by perspectival distortion using Lightroom's powerful Lens profile corrections

50 **Reveal more midtone detail and texture**
Reveal more texture and detail by selectively increasing midtone contrast using the Clarity slider

Teach yourself **Lightroom**
PHOTO-FIXING TOOLS

GET THE FILES HERE: http://bit.ly/tylr2016

BEFORE

AFTER

Remove lens distortion

Set up and apply lens profiles in Lightroom to automatically counteract lens-induced distortion

What you see with the naked eye isn't always what you get in a photo of the same location. When shooting with a wide-angle lens, the edges of the frame can become distorted, with horizontal and vertical lines appearing curved instead of straight. This barrel (outward) or pin-cushion (inward) distortion is especially noticeable when looking [at architectural photos that feature lots of straight lines.

Cheaper lenses can also add colour-related artefacts to a subject. When faced with a high-contrast edge (such as a white swan against a dark blue lake), you may find fringes of purple clinging to the edges of your subject. This ugly and distracting chromatic aberration is caused by the lens's inability to focus different wavelengths of light onto the same spot on your camera's sensor. Also, when you zoom out with your lens to capture a wider view of a landscape, the frame can become darker at the edges and corners. You'll also notice darker edge colours and tones when using a wider aperture. This vignetting effect occurs because different amounts of light are entering the edge of the lens compared to the centre, causing the corners of the image to become incorrectly exposed.

Fortunately, Lightroom's Lens Corrections panel has all the tools you need to counteract vignetting, barrel and pin-cushion distortion and chromatic aberration. Thanks to its collection of Lens Profiles, Lightroom can counteract these problems automatically, so that your photos will look more like the scene did when you saw it with the naked eye.

Teach yourself **Lightroom**

PHOTO-FIXING TOOLS

1. Add a grid
Bring TYLR16.dng into Lightroom's Library. To see if lines are distorted it can be handy to summon a grid. Go to View>Loupe Overlay and click Grid. Hold down Cmd/Ctrl and click and drag on the Size and Opacity options to customise your grid to suit the photo.

2. Read the metadata
Toggle open the Metadata panel. Here we can see that a 20-35mm f/2.8 lens was used to capture the photo. The focal length was 22mm. The camera was a Nikon D700. Lightroom can use this information to work out where the distortions will occur (and therefore counteract them).

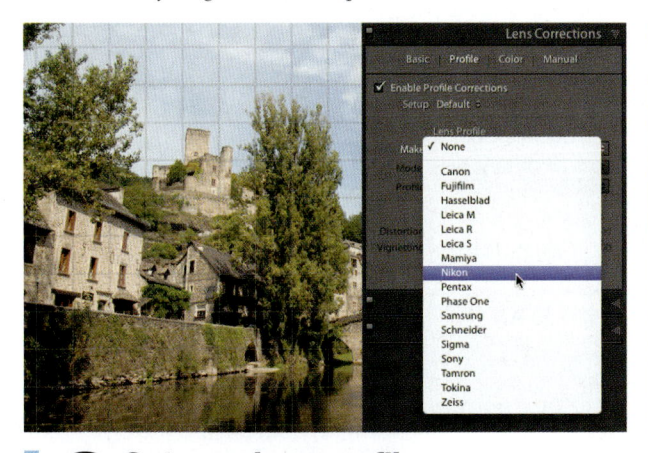

3. Set up a lens profile
Click the Develop module at the top. Toggle open Lens Corrections. You could go to Manual and drag the sliders to counteract distorted lines. For quicker, more precise corrections click Profile. Tick Enable Profile Corrections. In Lens Profile, click Make and choose Nikon.

4. Compare versions
The Lens Profile command reads the photo's metadata and then automatically selects the appropriate lens from Lightroom's list of lens profiles. Toggle the Enable Profile Corrections box on and off to see how the profile counteracts the original photo's distorted lines.

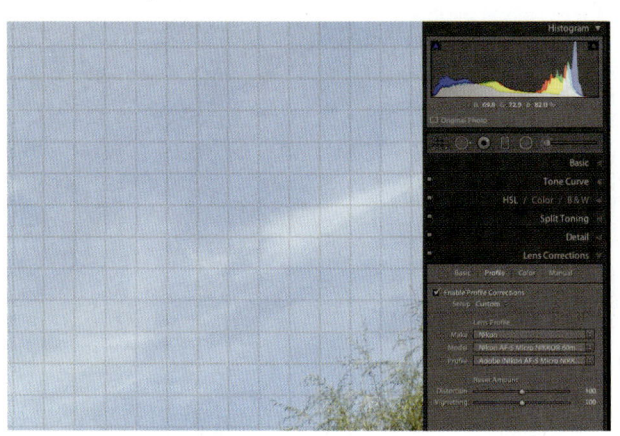

5. Remove the vignette
You can fine-tune the results achieved by dragging the Distortion slider, although the profile should do a good job. Toggle the correction on and off and you'll also notice that the original corners are slightly dark. The profile lightens them to match the rest of the scene.

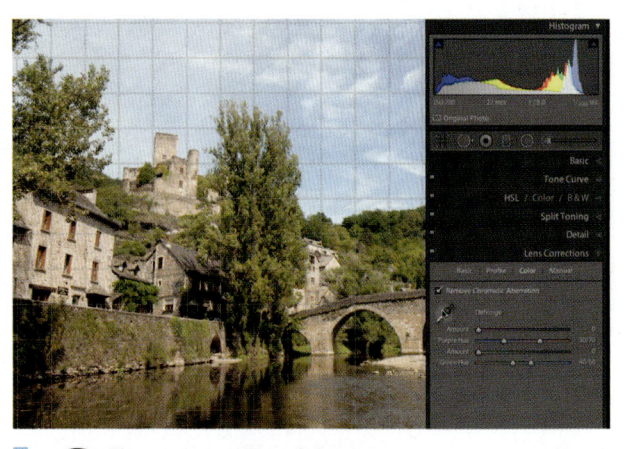

6. Remove the fringes
If you go to the Color tab you can turn on Remove Chromatic Aberration. This option counteracts any colour fringing around high-contrast objects with a click. You can fine-tune the results by using the eyedropper to sample unwanted fringe colours, then adjust the Amount slider.

Teach yourself **Lightroom**

PHOTO-FIXING TOOLS

GET THE FILES HERE: http://bit.ly/tylr2016

Get a better perspective

Straighten converging verticals caused by perspectival distortion using Lightroom's powerful Lens profile corrections

Most buildings have vertical walls at a 90-degree angle from the ground. However, in photographs the walls often appear wider apart at the ground, but tilt inwards towards the top. These converging vertical lines are created by perspectival distortion.

Tilting your camera to get the top of the building into the photo when shooting up close increases this distortion, as does using a wide-angle lens to fit the sides into the frame.

To avoid these converging vertical lines, you could try shooting your subject from a more distant (or higher) vantage point, so you no longer need to tilt the camera upwards. This will help the subject's walls run parallel with the edge of the frame. Alternatively, you could buy a specialist tilt-shift lens that counteracts the converging verticals as you shoot.

As you'll discover in our walkthrough, Lightroom provides several ways to straighten out converging verticals in its digital darkroom. The Lens Correction panel enables you to manually or automatically counteract perspectival distortion and make walls run parallel with the edges of the frame. In more recent versions of Lightroom you can adjust the parallax effect via the Transform panel in the Develop module. You can click and drag on the image to dictate particular uprights, which is very helpful for a quick edit.

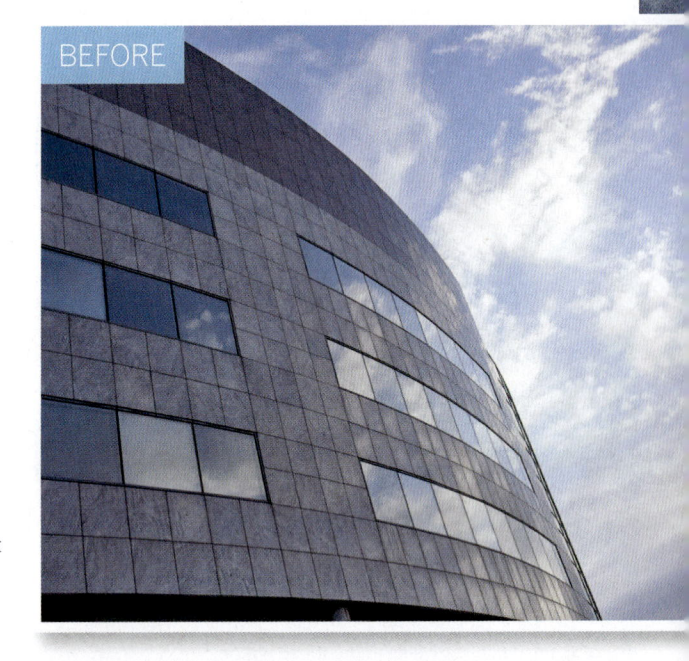

BEFORE

1 Crop the edges
Straighten your image first by cropping in to make it as upright as possible. You will lose part of your image, so if you're shooting first to manipulate the parallax effect shoot 10% wider than you normally would to counteract this. Now you can start to make manual adjustments.

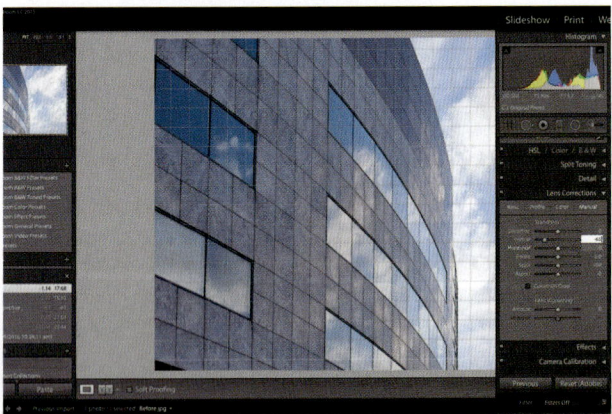

2 Make a manual correction
Take Building.jpg into the Develop module. Toggle open the Lens Corrections panel and click Manual. Drag Vertical to -65 to make the vertical lines run more parallel with the frame edges. If you do this step before the cropping you'll lose detail at the top and bottom of the frame.

Teach yourself **Lightroom**
PHOTO-FIXING TOOLS

AFTER

3 Make a profile
Click the Lens Corrections panel's Profile tab, then tick Enable Profile Corrections. Set Make to Nikon. This counteracts vignetting, subtle barrel or pin-cushion distortions and can help to fix perspectival distortion. Adjust the amount of Distortion and Vignetting using the sliders.

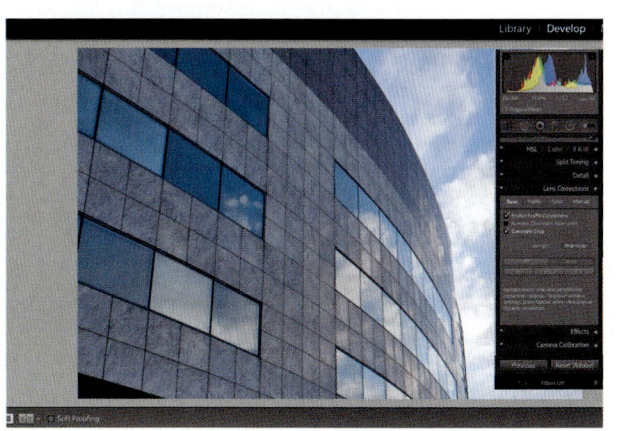

4 Automatic correction
If you can't get it right manually, go to the Lens Correction panel's Basic tab. Tick Constrain Crop to avoid white edges around the edge of your frame. Tick the Upright section's Auto button to automatically counteract tilted horizons and converging verticals.

Teach yourself Lightroom
PHOTO-FIXING TOOLS

GET THE FILES HERE: http://bit.ly/tylr2016

Reveal more midtone detail

Reveal more texture and detail by selectively increasing midtone contrast using the Clarity slider

At the start of this chapter we tackled a high-contrast scene that had both over- and under-exposed areas. We used the Develop module's Basic panel to independently adjust the shadows and the highlights to reveal detail where it was desired. The starting image on this page has some bright areas in the background windows, but the majority of the photo consists of dull shadows and murky midtones. The flat-lit printing press has lots of interesting textures and details, but in the unprocessed picture these areas of interest are lost in a muddy wash of dull brown midtones.

To reveal more detail in this scene we could have captured more tonal information in a series of bracketed exposures. However, Lightroom lacks Photoshop CC's ability to merge multiple photos as a single HDR (high dynamic range) composite that features detail in the shadows, midtones and highlights. We could also have zapped the scene with a burst of flash, but we chose to use the available mix of artificial tungsten and natural daylight instead because introducing a third light source would make the image unworkable.

Fortunately, our starting image is a raw file, so it already contains more tonal information than we can see when looking at the unprocessed version of the photo. In the walkthrough overleaf we'll show you how to lighten the midtones and increase their contrast to make finer features and textures stand out more effectively. We'll also show you how to reveal the scene's true colours by cooling down a warm colour cast and removing a slight magenta tint caused by an incorrect white balance setting.

AFTER

Teach yourself **Lightroom**

PHOTO-FIXING TOOLS

Teach yourself **Lightroom**
PHOTO-FIXING TOOLS

1. Examine the histogram
Import TYLR19.dng into Lightroom and take the unprocessed photo into the Develop module. From the spread of the histogram graph we can see that the photo has information throughout the entire tonal range. However, from the height of the graph we can see that the strongest tones are the shadows, followed by weaker midtones. We'll need to move (or remap) some of that shadow information into the midtone section to reveal missing detail.

2. Correct the colours
Drag the Temperature slider left to a value of 3418 to remove the photo's warm colour cast. To remove the hint of magenta, drag the Tint slider down from +30 to +22. This restores the metal printing press's greenish metallic hue and makes it look less brown. The cooler colours of the press now contrast with the warmer colours of the wooden bench. We now have more colour variety, which helps to differentiate the objects in the scene.

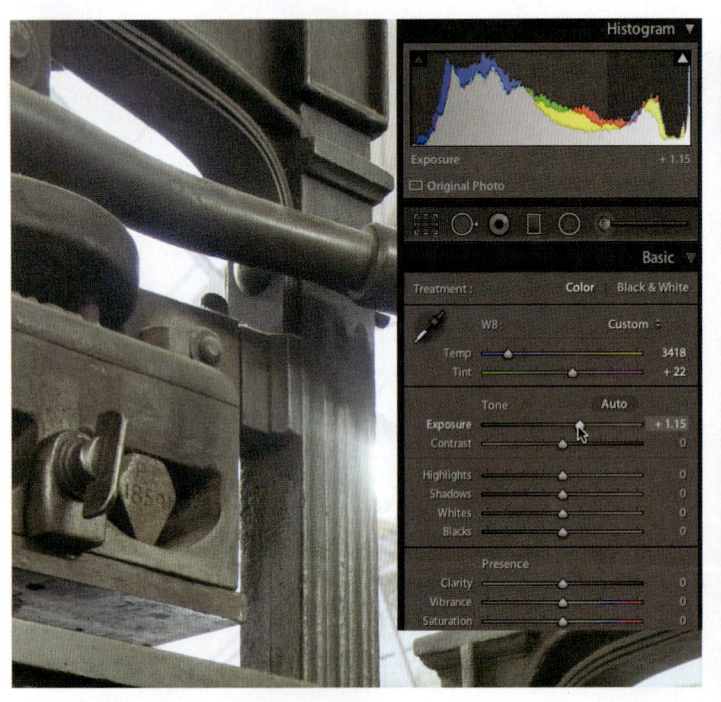

3. Increase the exposure
It's always a good plan to fix colour-related problems first, because this can change the tones in the image. This is why the white balance controls are at the top of the Basic panel. To reveal more detail, kick off by dragging the Exposure slider right to +1.15. The histogram graph will slide to the right, because some of the shadows become midtones. We can now see a healthier-looking histogram and more detail in the printing press.

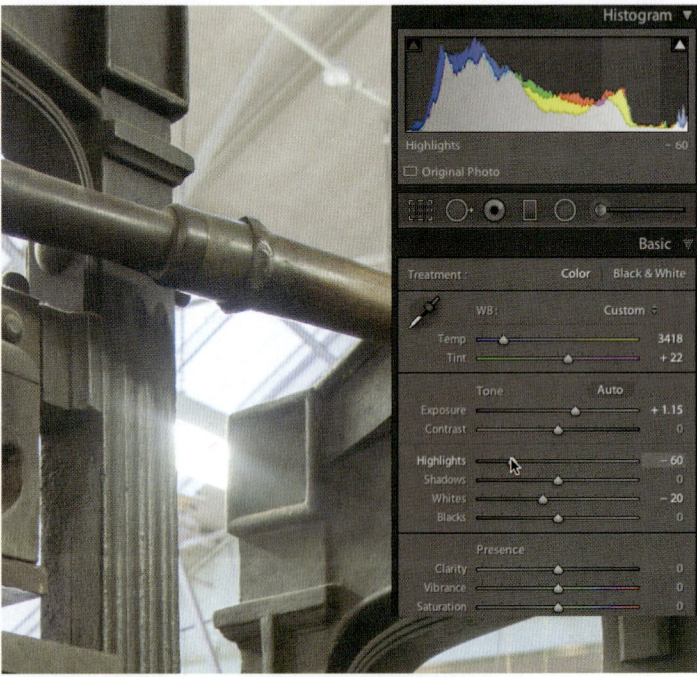

4. Remove the clipped areas
This global exposure change has blown out the highlights. Click the Highlight Clipping Warning icon at the top right of the histogram window. Over-exposed highlights will appear in patches of red. Move the cursor onto the far right of the histogram graph. The Whites sliders will become highlighted below. You can click and drag the graph to reduce the Whites to -20 (or drag directly on the Whites slider). Drag Highlights to -60 to remove the remaining red patches.

Teach yourself **Lightroom**

PHOTO-FIXING TOOLS

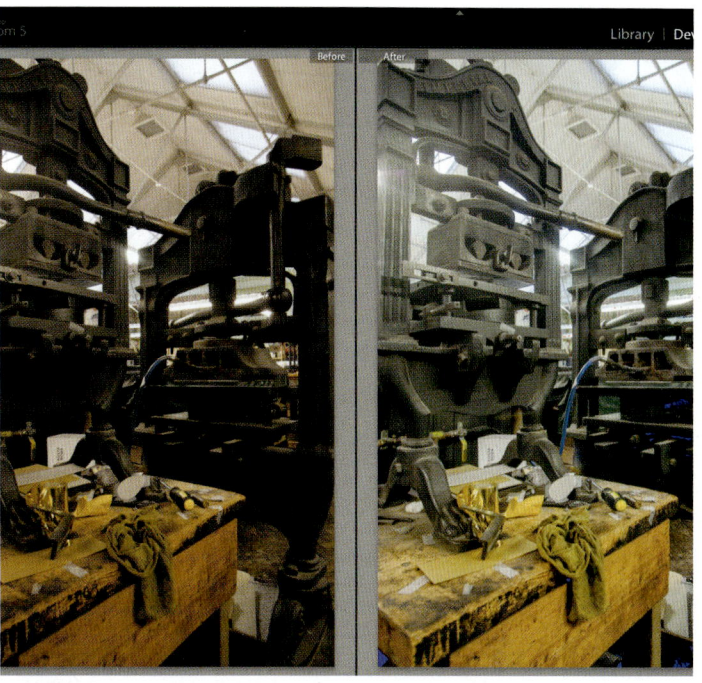

5 Improve the global contrast
To reveal even more midtone detail in the photo, drag the Shadows slider right to +72. This selectively lightens more of the shadows, and places them more towards the middle of the histogram window. You can now see even more detail in the printing press, but the photo lacks contrast and there aren't any strong blacks on the left of the histogram. To improve the photo's overall contrast, drag Blacks left to -65.

6 Compare before and after
Turn on the Shadow Clipping Warning (at the top left of the histogram window). A few blue patches will appear, indicating areas of pure black. It's good to have some black pixels in a photo, and because these areas don't contain any important detail we can leave them clipped. Click the Before and After icon at the bottom left to see how your processed picture is shaping up. You can now see much more detail in the midtones.

7 Increase the midtone contrast
You now have a healthier-looking histogram with more information in the midtones and some contrasting shadows and highlights. To tease out more midtone detail, drag the Clarity slider to +69. This increase in midtone contrast causes fine textures in the metal press and wooden bench to jump out. Zoom in to 1:1 magnification to compare the before version of the image with the after and see the revealed midtone detail.

8 Fine-tune the clipped patches
Go back to the Loupe view (by clicking its icon or pressing D). The edited photo will fill the workspace. The Clarity's slider's increase to midtone contrast may have caused a little clipping in the background's brighter areas, so drag the Whites slider left to -70 to remove some of these clipped patches. A few red patches can remain in the brightest windows, because these areas contain no important details.

Teach yourself **Lightroom**
SELECTIVE IMAGE ADJUSTMENTS

Teach yourself **Lightroom**
SELECTIVE IMAGE ADJUSTMENTS

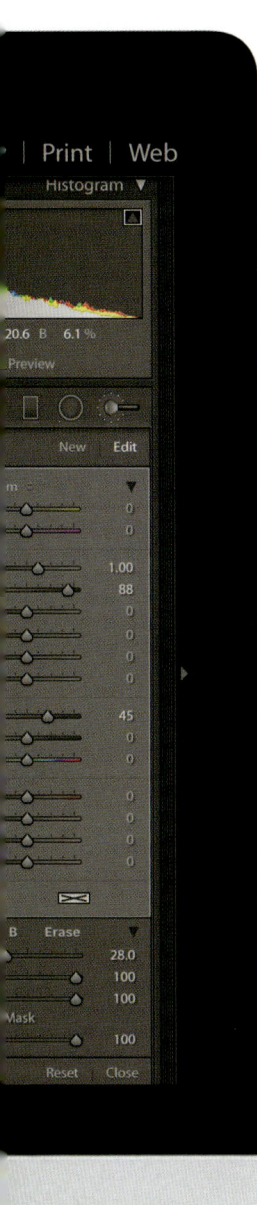

Selective adjustments

Edit selected parts of an image while leaving the rest of the image untouched

56 **Remove sensor spots in your images**
Use Lightroom's Spot Removal tool to remove camera sensor spots and other dust marks with a few clicks

60 **Dodge and burn in Lightroom**
Use the Adjustment brush to lighten and darken specific areas of a monochrome image non-destructively

64 **Editing in Lightroom with Auto Mask**
Identify areas to be selectively altered using the Adjustment brush's Auto Mask and Edit pins that enable fine-tuning

68 **Use the Graduated Filter tool**
Learn how to use one of the most useful tools for selective adjustments, and discover how to make essential edits to your landscapes

70 **Master the Radial Filter tool**
Learn how to draw the eye in towards your subjects by making circular adjustments to your images that create a subtle vignette

Teach yourself Lightroom
SELECTIVE IMAGE ADJUSTMENTS

GET THE FILES HERE: http://bit.ly/tylr2016

AFTER

Teach yourself **Lightroom**

SELECTIVE IMAGE ADJUSTMENTS

BEFORE

Remove sensor spots

Use Lightroom's Spot Removal tool to remove sensor spots and other dust marks with just a few clicks

Despite our best efforts to keep our lenses and cameras clean, we may find that some photos suffer from sensor spots. When you swap lenses over on location, tiny fragments of dirt and dust can enter the camera body and adhere to the sensor. These sensor spots can then show up in a photograph as small, soft, dark blobs. These blobs are particularly annoying because they appear in the same parts of every photo from a shoot. They are particularly noticeable in clean smooth areas such as skies (as you can see in the starting image).

Your DSLR camera may attempt to remove sensor spots by vibrating the sensor, but this automatic mechanical technique may not prove effective. For a more thorough clean you can activate the camera's sensor cleaning feature. This locks up the mirror and enables you to use a blower brush to dislodge particles of dust without damaging your camera's mechanisms. It's worth trying to nip sensor spots in the bud, because this will save you lots of time 'dust busting' in Lightroom's digital darkroom.

Even after cleaning your camera's sensor you may find that some sensor spots are still visible. In this instance you'll need to take the photo into Lightroom's Develop module and remove the blobs with a few clicks of the Spot Removal tool. This powerful tool samples clean sections of the photo that are adjacent to a sensor spot. The sampled section is then transplanted over the unwanted spot to hide it. The sampled pixels are seamlessly blended with their new surroundings to create an invisible edit, as you'll see overleaf.

Teach yourself Lightroom
SELECTIVE IMAGE ADJUSTMENTS

1 Import the image
Download our TYLR22.dng start image, go to File›Import Photos and Video, and bring the photo into Lightroom's Catalog. We've already cropped the photo and converted it into black and white, but it suffers from lots of undesirable sensor spots. Click Develop in the Library module to take it into that module. Go to the Navigator panel on the left and click 1:1. This gives us a closer look at the individual pixels.

2 Zoom and pan
Once you've zoomed in using the Navigator's presets, the Hand tool automatically becomes active. Click and drag the mouse to look at sections of the magnified photo. Dark sensor spots are clearly noticeable because they contrast with the light greys of the sky and the white foreground foam. Sensor spots are less noticeable when they overlap darker busier textured areas such as the harbour wall in this image.

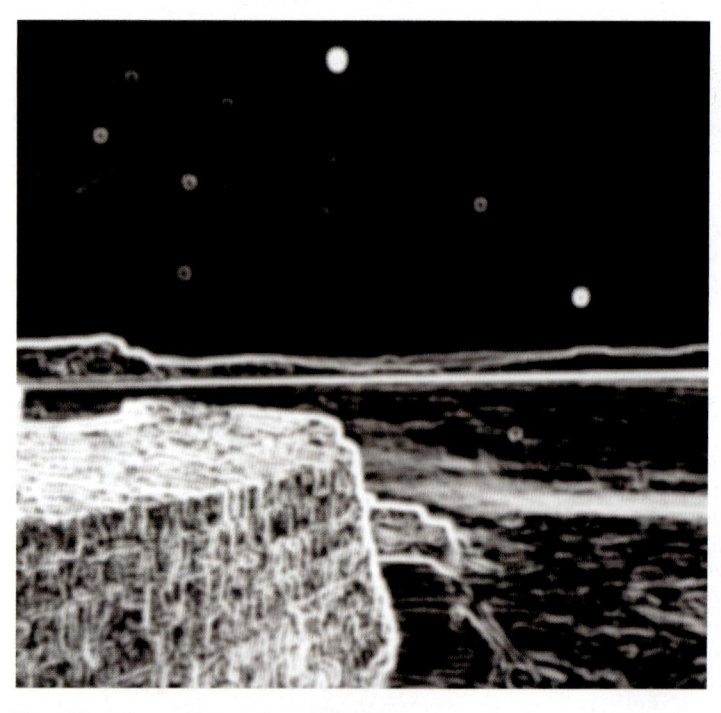

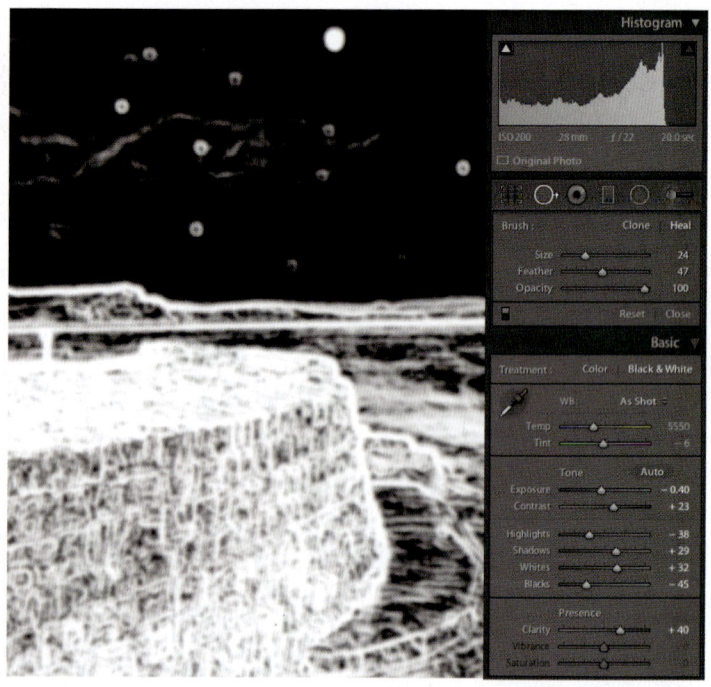

3 Visualise the spots
Go to the Develop module's toolbar (just below the histogram window) and click the Spot Removal tool icon. You can also summon this tool by pressing Q. Once you've activated the tool, go to the options bar below the photograph and tick the Visualize Spots box. This creates a greyscale preview of the photo. Prominent sensor spots show up as white blobs. More subtle spots appear as patches of grey.

4 Adjust the threshold
To discover the location of more subtle sensor spots, drag the Visualize Spots slider to the right. This increases the contrast of the spot visualisation threshold so that the faded grey spots become whiter and easier to locate. You'll also notice that the cursor changes to show the Spot Removal tool's target-shaped circular overlay. This overlay automatically appears when you move the cursor over the photo.

Teach yourself **Lightroom**
SELECTIVE IMAGE ADJUSTMENTS

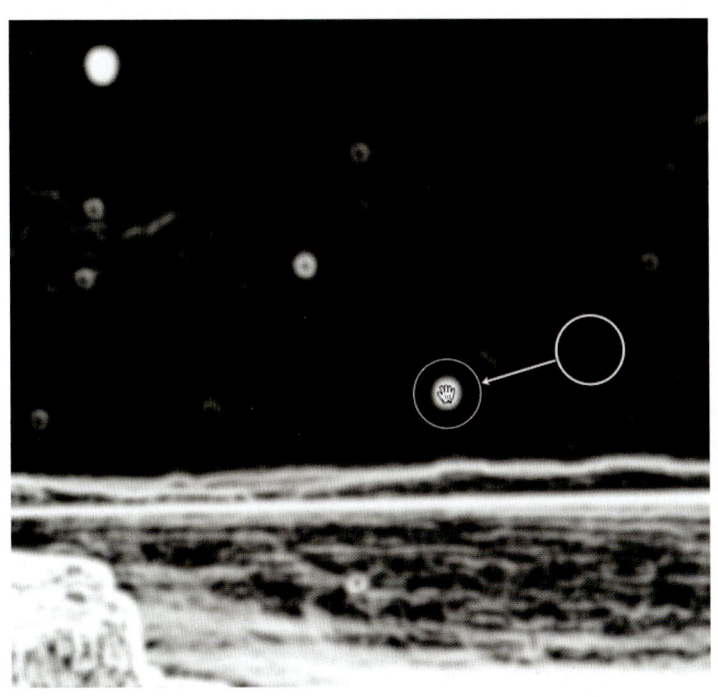

5 Set up the brush
Some of the sensor spots will be larger than the Spot Removal tool's circular overlay, so go to the section below the toolbar and increase the Size slider to 53. Set Feather to 47 for a soft edge. To seamlessly blend the clean patches of sampled sky with their new surroundings, click the Heal option rather than Clone. Healing produces an invisible edit that hides the spot more effectively. Place the overlay on a white spot and click.

6 Remove the spots
The Spot Removal tool will automatically sample an adjacent patch of clean sky and transplant it over the unwanted spot. An arrow indicates that we've taken a patch of clear sky from one circular overlay and placed it into another. The white Visualize Spot warning in the overlay will turn black to indicate that the targeted sensor spot is no longer visible. Clear Visualize Spots to see how effective the Spot Removal tool has been at hiding the targeted sensor spot.

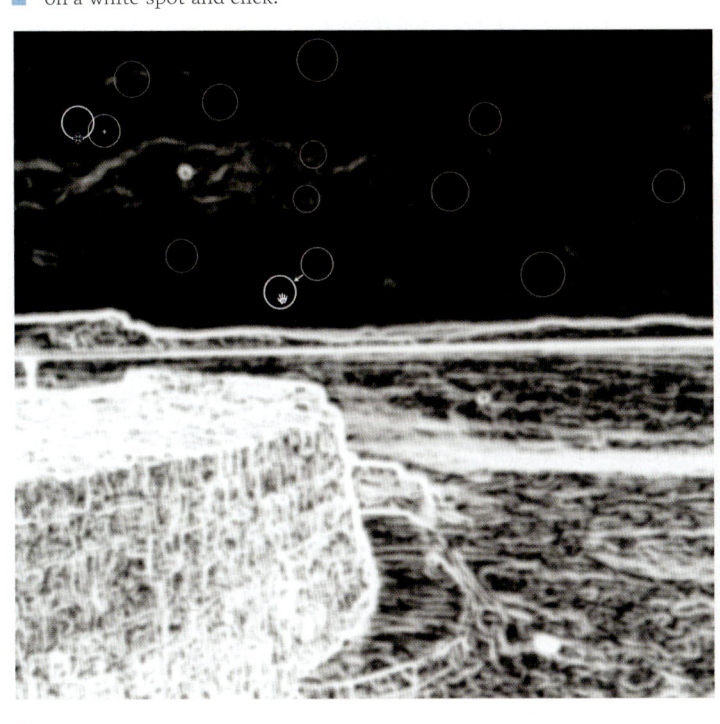

7 Fine-tune the overlays
Turn Visualize Spots back on. Click to heal other white and grey spots. To see all the Overlays you create, set the Tool Overlay drop-down menu to Always. You can drag inside an overlay to fine-tune its position. Drag the edge of an overlay to shrink or enlarge it. You can also tap the left square bracket to shrink the Spot Removal tool's overlay so that it can deal with smaller spots. Tapping the right square bracket enlarges the overlay.

8 Create irregular overlays
There's a stray hair on the left-hand side of the frame. Set Size to 29 and drag to paint the Spot Removal tool over the hair. This creates an irregular shaped overlay that's perfect for hiding the hair without altering too much of the surrounding sky. Continue clicking to place circular overlays over the remaining sensor spots. When hiding spots on the sea you may need to manually reposition the source overlay to sample and transplant similar pixels. Click Done to finish.

Teach yourself Lightroom

SELECTIVE IMAGE ADJUSTMENTS

GET THE FILES HERE: http://bit.ly/tylr2016

Dodge and burn

Use the Adjustment brush to lighten and darken specific areas of a monochrome image

When it comes to exposure, most photos will benefit with a few slider tweaks in Lightroom's Develop module to reveal more tonal detail. The Basic panel's Exposure slider makes global adjustments that evenly lighten or darken the tones in the photo. This slider works well if the entire image is slightly under- or over-exposed. However, if you use the Exposure slider to lighten under-exposed shadows, then you may blow out (or clip) correctly exposed highlights. If you try to claw back missing highlight detail with the Exposure slider, then you'll plunge the shadows into clipped darkness.

To help you make selective tonal adjustments you can use other sliders (such as Highlights and Shadows) to target and adjust specific tones, as we demonstrated at the start of chapter four. In that walkthrough we lightened a model's monochrome midtones to make her stand out from a room's gloomy shadows. The Highlights and Shadows sliders were able to target and adjust brighter and darker tones independently of each other, revealing detail where it was needed (and hiding distracting detail too).

The Highlights and Shadows sliders do a good job of enabling you to make selective tonal adjustments, but with some photos you'll need more tone tweaking precision. In this spread's starting image the camera has metered for the brighter sky, so we're losing detail in the silhouetted standing stones. We'll demonstrate how to use the Adjustment brush to make precise tonal adjustments to irregular shapes, so that you can achieve a more balanced exposure in a high-contrast scene. Thanks to the Adjustment brush you can dodge (lighten) or burn (darken) any area.

Teach yourself Lightroom
SELECTIVE IMAGE ADJUSTMENTS

BEFORE

Teach yourself **Lightroom**
SELECTIVE IMAGE ADJUSTMENTS

1. Brighten the image
Bring the TYLR22.dng starting image into Lightroom's Develop module. In the Basic panel, click the Black & White tab. This throws out all the colour information. At this stage the photo's murky shadows lack contrast and detail. Before we unleash the Adjustment brush, we can start to improve things with the sliders in the Basic panel. Drag Exposure up to +1.25. The histogram now shows more highlight information on the right.

2. Increase the contrast
Increase the Contrast slider to +62 for stronger shadows and highlights. The histogram's shadows and highlights look nice and strong, but there's a big gap in the middle, indicating a lack of midtone detail. To restore detail to the brightest highlights, drag Highlights down to -91. Push Whites to +26 for a wider range of tones. A good monochrome conversion should have some black blacks and white whites.

3. Increase the midtone contrast
To tease out missing midtone detail, drag the Clarity slider right to a value of +34. This increases midtone contrast and begins to reveal some of the texture in the stones. After using the sliders to improve the contrast and reveal more detail, we can move on to make more targeted and stronger tonal adjustments with the Adjustment brush. This will help you to selectively dodge the stones and burn the sky to reveal more detail.

4. Set the Adjustment brush
Click the Adjustment brush icon in the toolbar under the histogram window (or press K to activate it). A new panel of sliders will appear immediately below. Click the Effect menu at the top of this panel and choose Dodge (Lighten). This adjustment preset automatically sets the Exposure slider to 0.25 (and the other sliders to 0. Manually increase Exposure to 0.72. Set the brush Size to 15 and Feather to 50.

Teach yourself Lightroom
SELECTIVE IMAGE ADJUSTMENTS

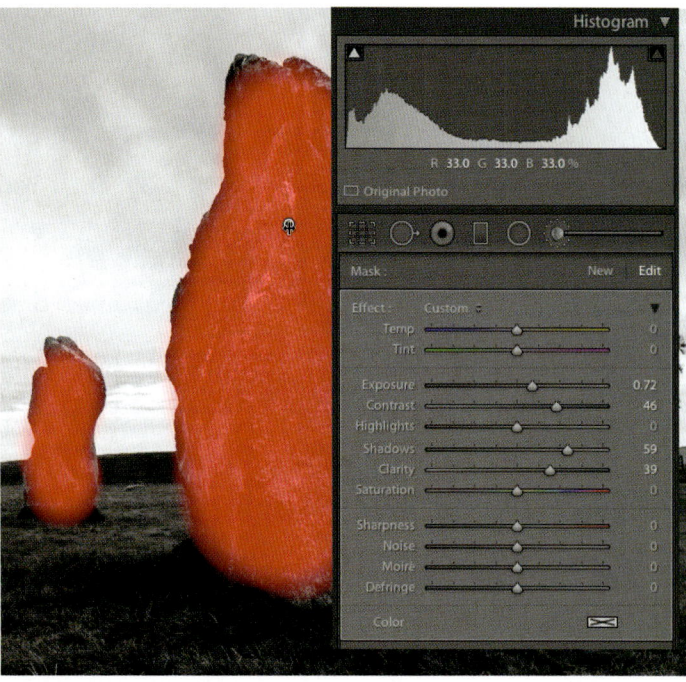

5 Fine-tune the adjustments
Click to place an Edit pin on the largest standing stone. Paint the Adjustment brush over the stone to selectively lighten it using Exposure. You can fine-tune the effect of the Adjustment brush after you've painted it in. Set Contrast to 46. Reveal detail in under-exposed areas by increasing Shadows to 59. Tease out more detail by pushing Clarity up to 39. This helps the stony, mossy texture to stand out.

6 Create a mask
Tap the left square bracket key a few times to drop the brush tip Size down to 9 (or drag the Size slider left). You can then paint with the Adjustment brush over the background standing stones using the same settings and lighten them up. To see which areas a particular Edit Pin is controlling, move the mouse onto the pin. A red mask will appear over the edited areas. We'll look at ways to use masks more effectively overleaf.

7 Set a new pin
The dodged standing stones have much more impact. To restore detail to the clouds we need to make a separate selective adjustment. In the mask section, click New. In the Effect section, click the drop-down menu and choose Burn (Darken). Drag the Exposure slider left to -0.74. To build up the adjustment in more controllable incremental strokes, go to the Brush section and set Flow to 27. Click to place a pin on the sky.

8 Add impact to the clouds
Tap the right square bracket key a few times to increase Size to 15, then paint over the clouds to darken their fine midtones. Avoid painting over the correctly exposed stones. To give the clouds even more impact, push Contrast to 45, Shadows to -78 and Clarity to 58. This increase in midtone contrast teases out the fine textures and details, and adds mood to the composition. Drop Exposure to -0.89 for more impact.

Teach yourself Lightroom
SELECTIVE IMAGE ADJUSTMENTS

GET THE FILES HERE: http://bit.ly/tylr2016

Editing with Auto Mask

Identify select areas to alter using the Adjustment brush's Auto Mask

On the previous spread we demonstrated how to use the Adjustment brush to dodge the dark standing stones, while burning more detail into the brighter sky. By altering the size, softness and flow of the Adjustment brush, you can target and tweak the tones of specific objects with precision. When you click a photo with the Adjustment brush you place an Edit pin. This pin records the position and strength of all the tonal adjustments you make. Thanks to Edit pins you can make multiple adjustments to a photo and then click a pin to fine-tune its effect at any time in the future. If you click a pin and drag it to the right, you can increase the value setting of each associated slider to, say, brighten the image more or boost the contrast more. Drag left on a pin to reduce the slider settings.

In step six on the previous spread we introduced masks. By moving the mouse over a particular pin, you can see a red mask overlay that indicates which areas are being adjusted by that pin. The Adjustment brush strokes we used to lighten the stones were soft and imprecise, so the brush tip could stray over the background and lighten the sky or ground. In the following walkthrough we'll show you how to use Auto Masking to dodge and burn more precisely, to lighten the circular light orbs without blowing out the background details, and deepen the cavernous shadows without darkening the surrounding moss.

Teach yourself **Lightroom**
SELECTIVE IMAGE ADJUSTMENTS

Teach yourself Lightroom
SELECTIVE IMAGE ADJUSTMENTS

GET THE FILES HERE: http://bit.ly/tylr2016

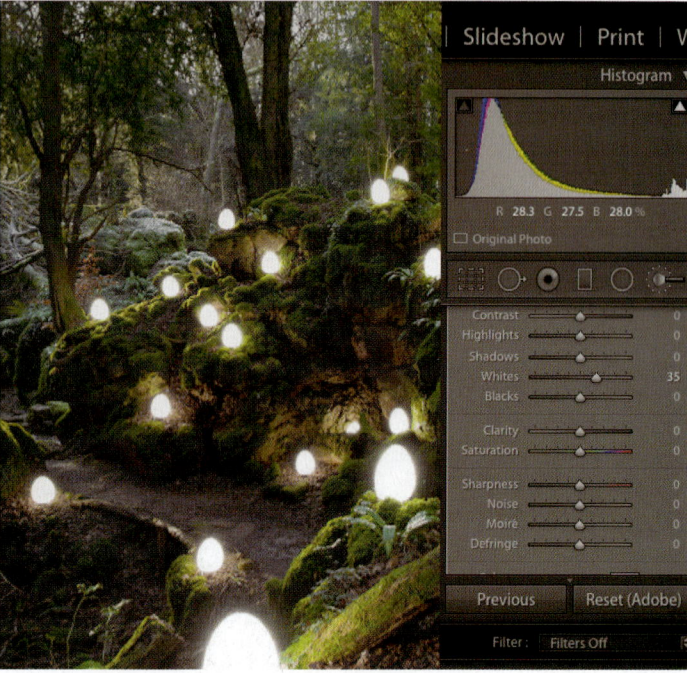

1 Improve the tonal range
Download Easter_egg_hunt.jpg to your computer, import it into Lightroom's Library and click the Develop module. This low-contrast starting image lacks punch in the shadows and the highlights. In the histogram window you can see that the black could be darker, because the graph doesn't quite stretch to the far left. Drag the Blacks slider left for more dramatic and darker shadows. Boost the Whites a touch to brighten the light orbs.

2 Adjust the brush attributes
To brighten the light orbs in the forest, first press K to select the Adjustment brush. Then, boost the Whites setting up to around +35 and set your brush size to 20. Next, scroll down and change your brush settings. Increase Feather and Flow to around +50 each. The orbs naturally have a slight glow so having a soft, flowing brush will help you to mimic the natural glow of the eggs scattered on the forest floor.

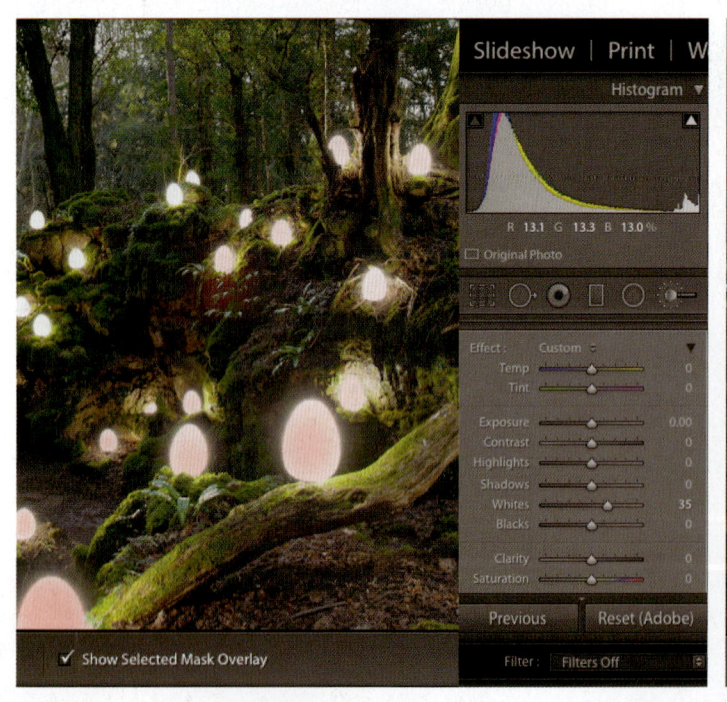

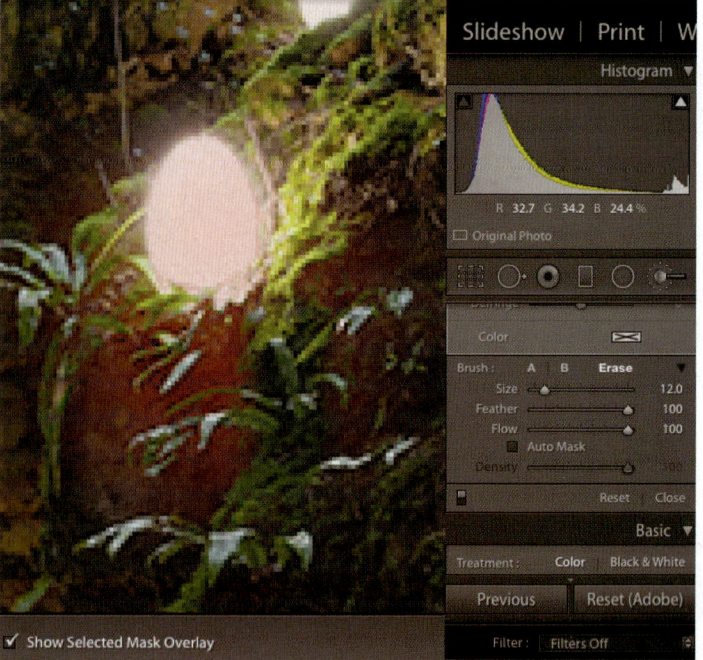

3 Activate the mask overlay
If you hover the cursor over the Edit pin, you'll see a red mask overlay indicating the already edited areas. You can turn the mask overlay on permanently by ticking the Show Selected Mask Overlay box at the bottom-left of the photograph. Some of the Adjustment brush's strokes may have strayed over the lighter background due to the different sizes of the eggs. Fortunately, you can tidy up the mask for a more precise adjustment after the fact.

4 Erase the problematic areas
Go to the Brush section on the right and click Erase, (you can also do this temporarily by holding the Alt key down on your keyboard). This gives you a new cursor with a minus icon inside it. Set Size to 12 and change Feather and Flow to 100 for a more accurate brush edge. Auto Mask Box enables the Erase brush to understand changes in contrast. There are sections that need removing completely though, so untick this while using the Erase brush.

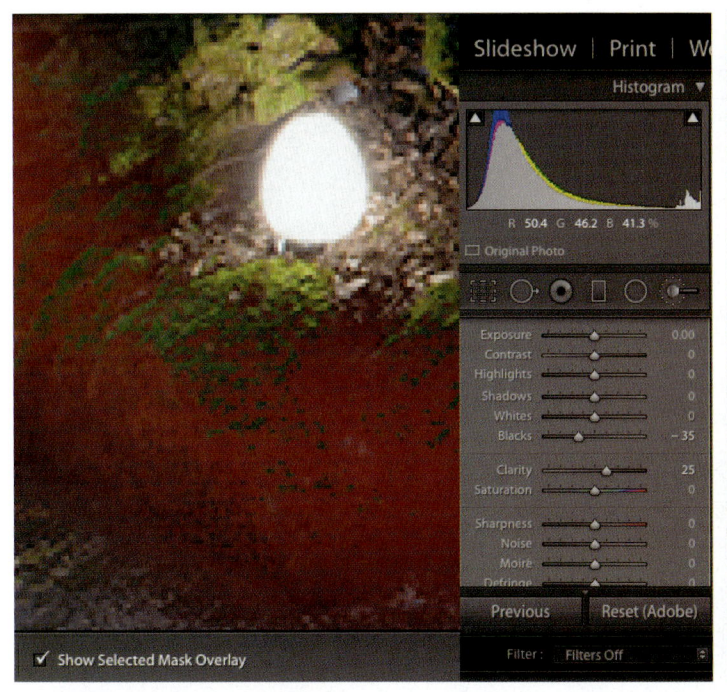

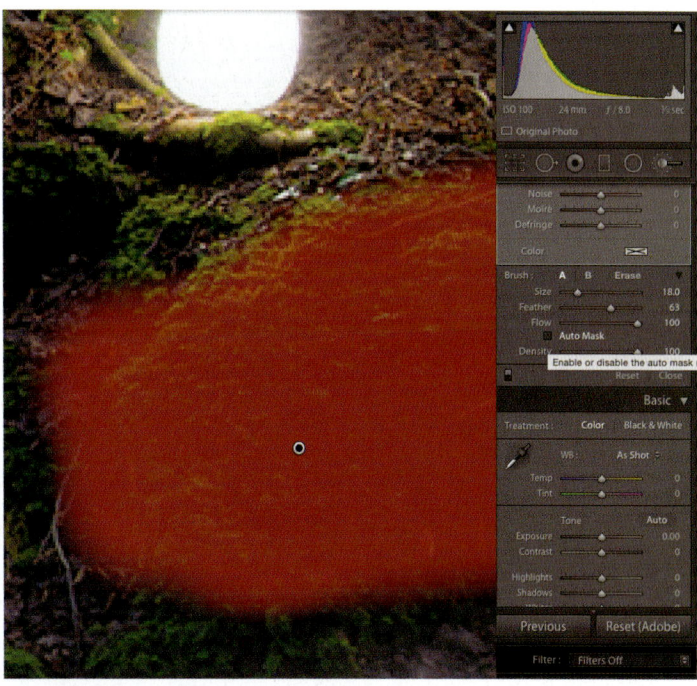

Teach yourself **Lightroom**
SELECTIVE IMAGE ADJUSTMENTS

5 Use Auto Mask
The Auto Mask feature will also help when making additional selections with the Adjustment brush. In the Brush section click back on A to access your original Adjustment brush settings. Tick the Auto Mask box. Now click 'New' at the top of the Brush panel, to make a new brush pin for us to make different adjustments. Set Blacks to -35 and Clarity to 25. Paint over the darker shadows on the image to deepen them further and increase mid-tone contrast for an added boost.

6 Turn off the mask
Auto Mask should avoid selecting lighter background details, so you won't need to do too much painting. Keep Auto Mask active as you paint around any remaining shadows to darken the cracks and crevices while leaving the moss and light orbs untouched. Turn off Auto Mask when painting whole areas of shadow that have no moss or light orbs to produce a more even-looking tonal adjustment. Turn off the mask overlay to see the lightened tones of the rocks.

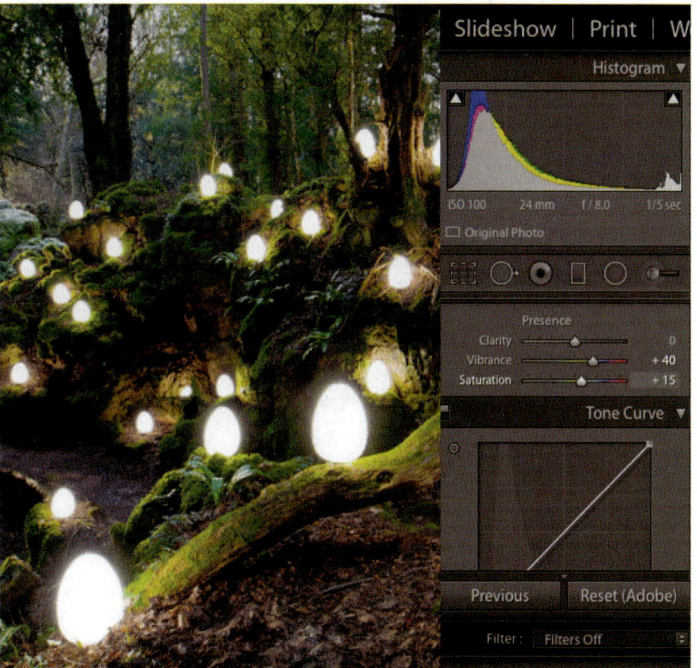

7 Remove the distractions
Now that you've darkened the forest floor shadows and made the light orbs are glow a brilliant white, check that the rest of the image has no distractions. The floor is covered in leaf litter and sticks, and at the bottom of the frame is a bright green leaf. It distracts from the warm tree trunk, so select the Spot Removal brush with Q and paint over the leaf. The brush automatically chooses a suitable place to clone from so the leaf disappears.

8 Add the finishing touches
Now that the whites are bright and the shadows deep, it's time to add a final polish to the overall image. Deselect your Spot Removal brush by pressing Q again, and this time scroll up to the Vibrance and Saturation section under the Basic tab in the Develop module. Increase Vibrance to +40 to boost the more muted colours until they're nearly as strong as the brighter colours. Then add +15 Saturation to give the whole scene some pizzazz.

Teach yourself Lightroom
SELECTIVE IMAGE ADJUSTMENTS

GET THE FILES HERE: http://bit.ly/tylr2016

BEFORE

The Graduated Filter tool

Learn how to use one of the most useful tools for selective adjustments, and how to make essential edits to your landscapes

The Graduated Filter is one of the most useful tools that Lightroom has on offer. It works by creating a gradated tonal blend between any two points in an image. The area behind your first point is completely affected, and beyond your second point remains unaffected, with a blend in between. What makes it such a powerful tool is the fact that, just like the Radial Filter and Adjustment Brush tools, you have complete control both over the area it affects, and the type of tonal changes that you'd like to make in that area. You can click and drag a line in any direction to create your blend, but the classic example of a graduated filter in action is to darken a sky in a landscape. One problem that landscape photographers often encounter is that skies are brighter than the land, so if you expose for the land, then the sky comes out too bright, or sometimes even completely white with no detail. But much like traditional lens-based graduated neutral density filters, you can use the Lightroom tool to darken a sky for a more even exposure. However, the control goes way beyond basic exposure changes: you can add contrast, boost colour and brush out areas you want to exclude from the effect. Here's how it works…

Teach yourself Lightroom

SELECTIVE IMAGE ADJUSTMENTS

1 Dial in some under-exposure
Bring your chosen image into Lightroom's Develop module, then grab the Graduated Filter tool from the toolbar at the top right of the interface. Notice that when you click the tool, a set of sliders appears directly below it. These closely resemble the Basic panel sliders, but the difference is that they work in combination with the tool. Set Exposure to -1.17.

2 Drag a gradient
Click and drag down from the top of the image to darken the sky. You can hold down Shift as you drag if you want to keep the lines perfectly horizontal. If you want to tweak the gradient, drag the upper or lower lines to change the width of the blend (hold down Alt and drag to move both at once) or drag the centre line to adjust the angle.

3 Add more gradients
You can add as many gradients as you like. Simply click and drag elsewhere in the image to make another, then adjust the sliders below the toolbar to change how the area looks. The tool will remember previous settings, but you can quickly reset everything to default values by double-clicking Effect, or by double-clicking on individual sliders to reset them.

4 Erase the mask
The gradient over the sky here is also darkening the lighthouse, but you can exclude it from the effect. Click Brush at the top right (next to New and Mask), then go to the Brush settings below the tool's tonal sliders. Choose A or B to add to the effect, and Erase to exclude, then paint over the area. Press O to toggle the mask overlay so you can see what's affected. ∎

Teach yourself Lightroom
SELECTIVE IMAGE ADJUSTMENTS

GET THE FILES HERE: http://bit.ly/tylr2016

The Radial Filter

Learn how to draw the eye towards your subject by making circular adjustments to your images

Just like the two other selective tools — the Graduated Filter and Adjustment Brush — the Radial Filter tool enables you to alter the tones in parts of an image. The difference is in the way you define the area you'd like to change. While the Adjustment Brush works by painting and the Graduated Filter works by creating a straight-lined blend, the Radial Filter enables you to create a circular blend between the area affected by the tonal changes, and the area that remains unaffected. As such it's very useful for creating subtle vignettes that draw the eye towards your subject and away from distracting edge details, or for softening parts of an image you want to de-emphasise.

Here we've used the tool for two very different tasks. For the first landscape image, you can drag a circle to darken down the corners of the frame and lead the eye towards the distant lighthouse. And for the second image, you can add two circular adjustments — one to claw back blown-out detail in the arm, and the other to add contrast to the face.

BEFORE

1 Drag a circle
Bring the starting image into the Develop module and grab the Radial Filter tool from the toolbar at the top-right. This opens up a set of sliders below that can be used to dial in tonal changes for the tool. Set Exposure to -0.65, then drag a circle over the lighthouse to subtly darken the corners of the image.

2 Add a vignette
If the circle isn't perfectly positioned, drag the pin to move it. When making a circle there are a few shortcuts that can help. Hold down Shift for a perfect circle, and hold down Alt to make the start point the corner of the circle. If you want to snap the circle to the edges of the image for a vignette, hold down Cmd/Ctrl and double-click with the tool.

Teach yourself Lightroom

SELECTIVE IMAGE ADJUSTMENTS

AFTER

3 Erase parts of the mask
Move on to the portrait image. Drag a fresh circle over the face, then dial in Exposure -0.49, Highlights -19. This improves the elbow, but makes the hand too dark. Click Brush at the top-right, then go to the Brush settings below the tonal sliders and click Erase. Press O to toggle the mask overlay on, and paint to erase the mask over the hand.

4 Change inside the circle
You can also make circular adjustments that affect the inside rather than the outside of the circle. Hold down Cmd/Ctrl+Alt, and drag the pin on the face to make a copy of the original circle. Check Invert Mask below the tonal sliders, then double-click Effect to reset them all and dial in Contrast +49 to add punch to the face.

Teach yourself **Lightroom**
SPECIAL EFFECTS

Teach yourself **Lightroom**

SPECIAL EFFECTS

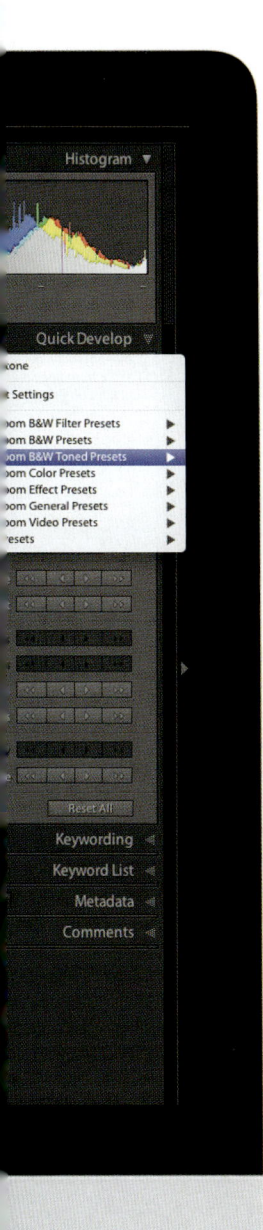

Special effects

Improve your photos fast with easy-to-apply special effects in Lightroom

74 **Master the HSL panel in Lightroom**
Enhance a photo by targeting specific colours and changing their hue, saturation and luminance values

78 **Make better black and white images**
Selectively lighten or darken greyscale tones in a mono conversion based on the original colours of an image

82 **Create high dynamic range images**
Discover how to combine several exposures into a single, highly detailed image using Lightroom's powerful HDR Merge feature

73

Teach yourself Lightroom
SPECIAL EFFECTS

GET THE FILES HERE: http://bit.ly/tylr2016

Master the HSL panel

Enhance a photo by targeting specific colours and changing their hue, saturation and luminance values

Certain colours are weaker in some photos than in others. If you boost the overall colour intensity of the image using the Basic panel's Saturation slider, then weaker colours will have more impact, but the stronger ones may become over-saturated. This can lead to colours that won't print correctly because they'll be clipped and lacking in detail. In addition, a model's skin tones may become too orange and garish. The Vibrance slider is designed to solve this problem by enabling you to increase the saturation of weaker colours without over-saturating the stronger ones. It also tends to boost natural landscape colours such as blues and greens more strongly than oranges, which helps you to avoid creating over-saturated skin tones.

If you venture beyond the Basic panel you'll find some alternative colour-tweaking tools that can help you enhance a problematic photo. In our unprocessed starting image the sunset is slightly blown out, and we're losing the vibrant colours that we saw on location. The hills and shoreline are under-exposed, so we can't see the textures and colours in these dark areas. The HSL (Hue, Saturation, Lightness) panel offers a powerful and effective way to target specific colours (such as the oranges and reds of the sunset) and change not only their colour intensity but also their lightness (luminance). This enables you to lighten and boost specific regions based on their colours, as you'll see in the walkthrough overleaf...

Teach yourself **Lightroom**
SPECIAL EFFECTS

BEFORE

Teach yourself Lightroom

SPECIAL EFFECTS

1. Use Solo Mode
Import TYLR17.dng into Lightroom and then take it into the Develop module. Here we'll use the Basic and HSL panels to darken the photo's blown-out highlights, lighten its under-exposed shadows and boost weak colours. You'll be jumping from one panel to another, so right-click any panel and choose Solo Mode. This means that when you click to open one panel, the previous one will close. This creates a tidier, cleaner workspace.

2. Lighten the shadows
Open the Basic panel. Here we can make some global adjustments to the photograph and then refine them using the HSL panel. Drag the Shadows slider right to a value of +47. This targets and lightens the under-exposed landscape's shadows without over-exposing the brighter sky's highlights. To selectively darken the sky we can use the Graduated Filter that we introduced in the previous chapter.

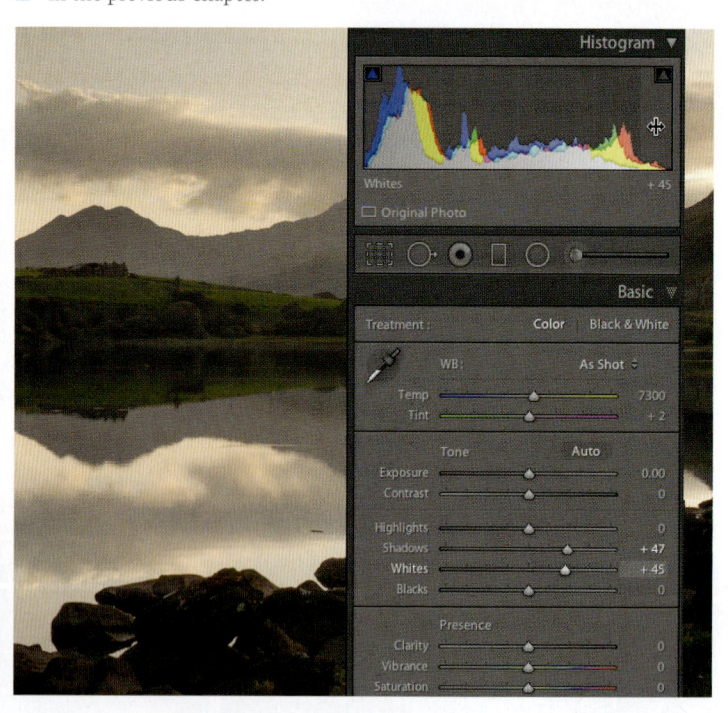

3. Darken the sky
Click the Graduated Filter icon in the toolbar below the histogram window. Choose Exposure from the Effect menu. Set the Exposure slider to -1.39. The other sliders should be set to 0. Click at the top of the frame and drag down to the horizon. This creates a gentle gradated tonal adjustment that darkens the top of the frame and reveals more colour and texture detail in the blown-out clouds. Click Done.

4. Expand the tonal range
By looking at the histogram we can see that the graph doesn't quite stretch to the far right. This indicates that the photo lacks strong highlight information. Click the cursor on the far right of the histogram window and the Whites slider will automatically become highlighted. Click and drag right to increase the Whites slider setting to +45. You now have a wider ranger of tones with some dark shadows and bright highlights.

Teach yourself Lightroom

SPECIAL EFFECTS

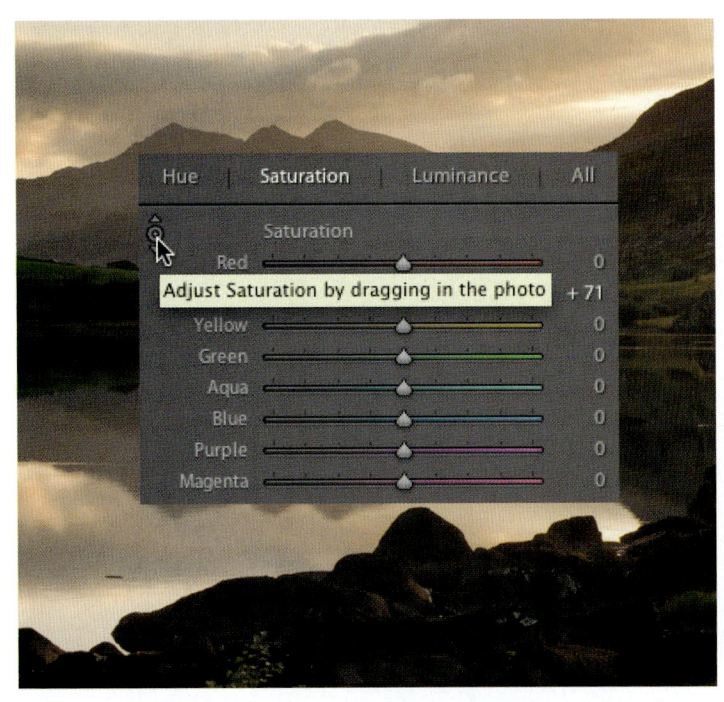

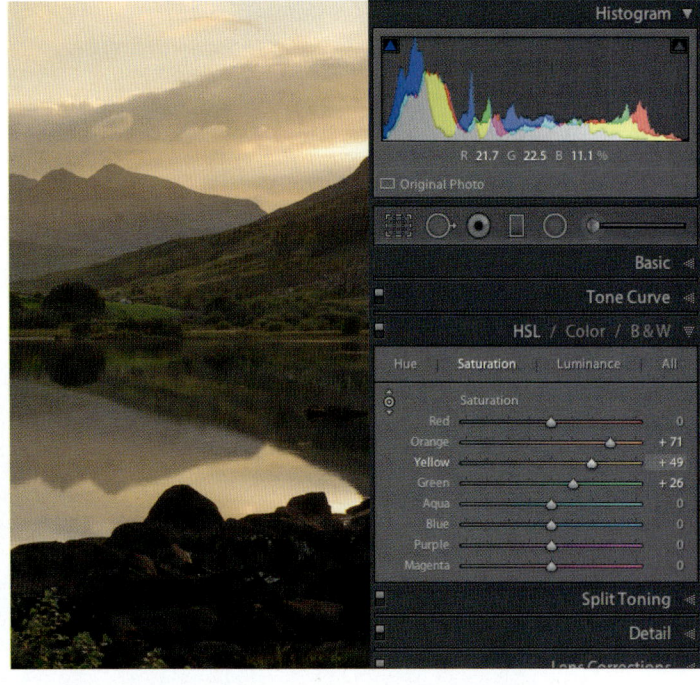

5 Selectively boost the colours
To refine colours and tones in selected parts of the photograph, click the HSL panel, and then click on the Saturation tab. By dragging the Orange slider to +71 you can boost the colours of the sunset without altering the colours in the rest of the photo. If you're not sure which colours are present in a darker area, click the Targeted Adjustment tool icon at the top left of the Saturation tab, then click a green hill to target it.

6 Make targeted adjustments
Click and drag upwards with the targeted Adjustment tool. This tool will sample the colours in that specific area and increase their saturation. Although the hill looks green to the naked eye, there are lots of yellows in the sampled area. This causes the Green and Yellow slider to move to the right. Stop dragging upwards when Yellow is at +49 and Green is around a value of +26. You can manually tweak individual sliders too.

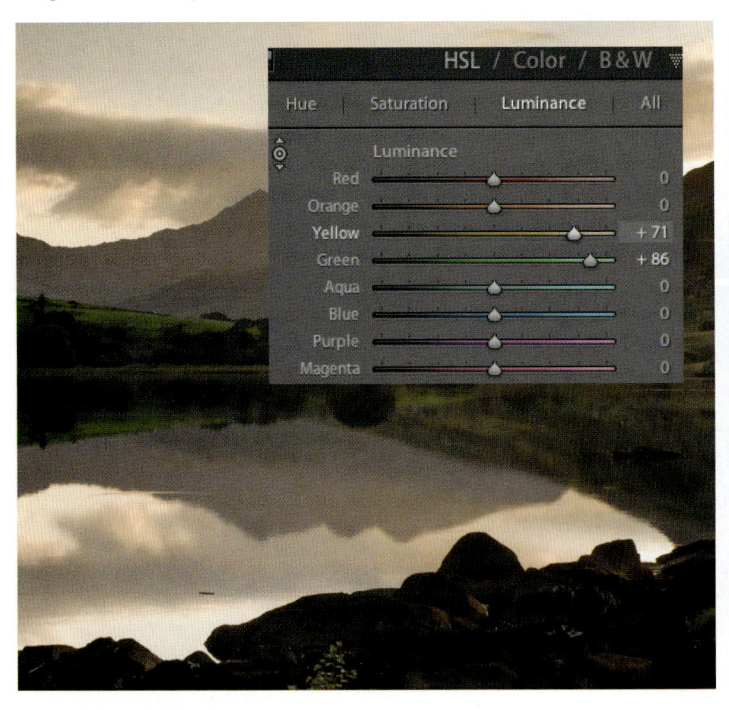

7 Lighten specific colours
The green hills are more saturated, but they're still lost in the shadows. To selectively lighten them, click the Luminance tab. Click the Targeted Adjustment tool on a hill to sample it and drag upwards to lighten the sampled colours. Let the Yellows shoot up to a brighter +71. The sampled Greens will increase by a lesser increment. Manually drag the Green slider up to +86 to lighten them a bit more.

8 Fine-tune the highlights
By boosting the Saturation and Luminance of the Yellows and Greens we can see more of the scene's colour, texture and detail. However, the lighter yellows have blown out the sky a little, so click the Graduated Filter icon, click its grey Edit pin to reactivate the adjustable sliders associated with that adjustment, and drag the Exposure slider left to a darker -1.93 to reveal a little more detail in the sky's brightest highlights. ∎

Teach yourself Lightroom
SPECIAL EFFECTS

GET THE FILES HERE: http://bit.ly/tylr2016

Teach yourself Lightroom
SPECIAL EFFECTS

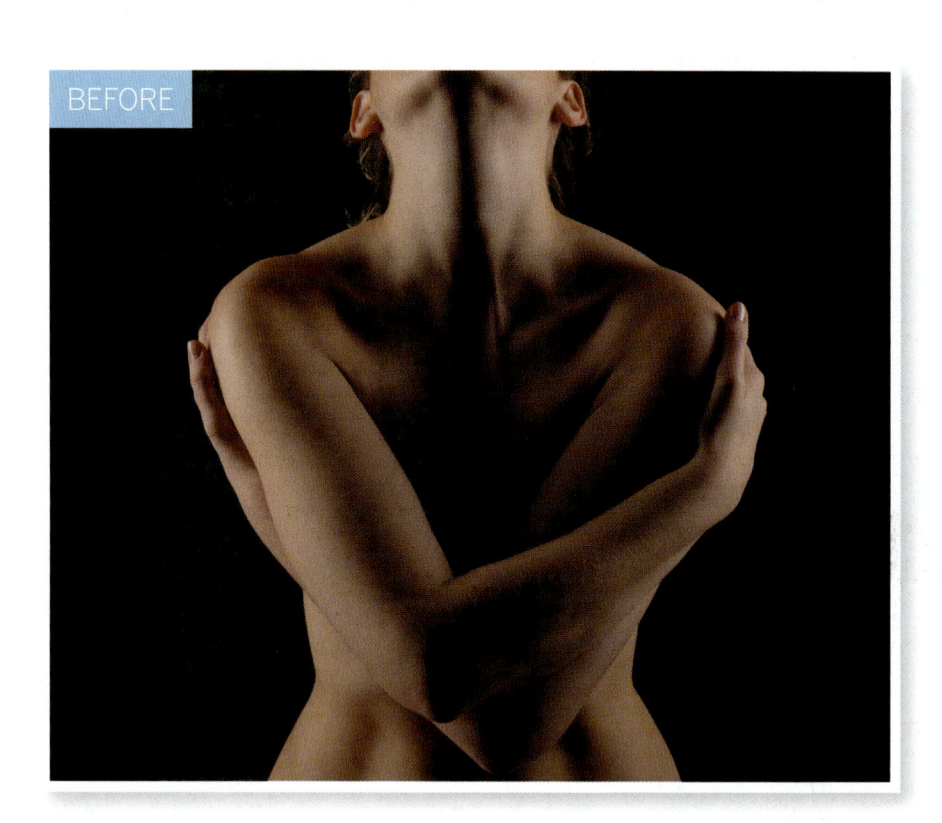

BEFORE

Better black and white images
Selectively lighten or darken greyscale tones in a mono conversion based on the original colours of the image

When shooting with black-and-white film, pre-digital photographers could also place a coloured filter over the camera's lens to produce more striking monochrome pictures. Different coloured filters would lighten or darken the greyscale tones of specific objects in the scene. For example, if you placed a red filter over the lens it would make blue skies become dark greys in the monochrome print, enabling lighter clouds to pop out in contrast.

Many digital cameras enable you to use a Monochrome picture style setting to produce a black-and-white photo in camera. You can even set the camera to apply colour filters that help lighten or darken greyscale tones in particular areas, such as blue skies or green fields. This in-camera approach can be very hit and miss, so we'll show you how to take more control of greyscale tones in your monochrome conversions in Lightroom.

There are several ways to create a monochrome image in Lightroom. If you simply throw away a photo's colour information by setting Saturation to 0, you'll get an instant mono conversion, but you risk producing a drab wash of greyscale tones that swallow up your subject's features, especially if they consist mostly of midtones. An eye-catching monochrome photo should have a wide range of tones, from black shadows through greyscale midtones to white highlights. A stronger tonal contrast helps to bring out a model's shape and contours, as you can see in our after image.

In the walkthrough overleaf we'll demonstrate how to use the B&W panel to selectively lighten or darken greyscale tones.

AFTER

Teach yourself Lightroom

SPECIAL EFFECTS

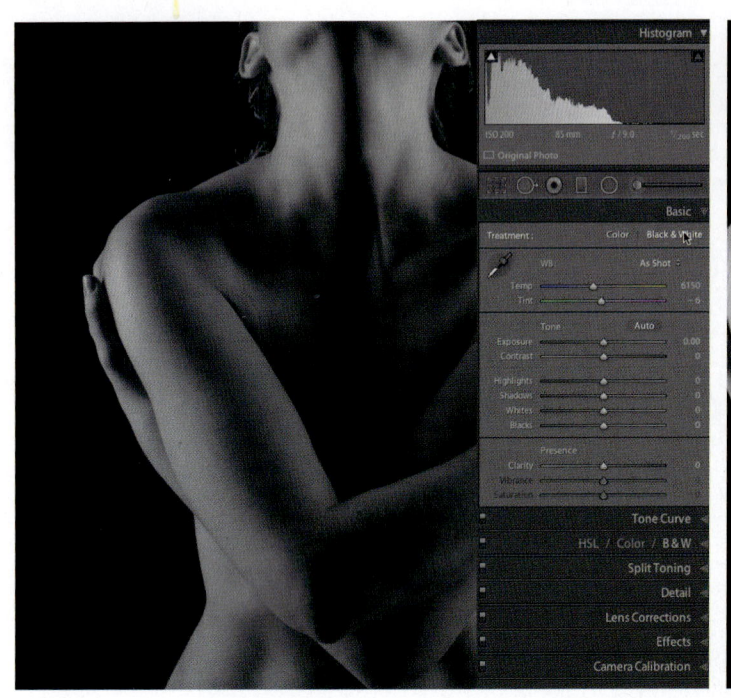

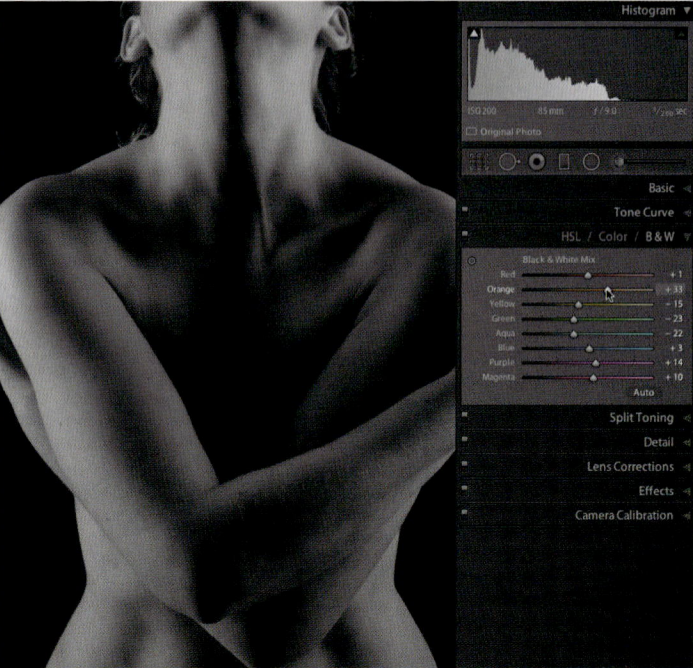

1 Convert to mono
Import TYLR01.dng into Lightroom, and take the photo into the Develop module. Go to the Basic panel and click on the Black & White tab. This desaturates the photo, but the result is a mass of murky midtones and shadows. If you look at the histogram window, you can see that the mono conversion lacks highlight detail. It also looks under-exposed to the naked eye. We need to improve the photo's contrast.

2 Lighten the colours
Scroll down and click the B&W panel. When you convert a photo to monochrome, Lightroom automatically adjusts the position of the Black & White mix sliders to try to get an effective black-and-white photo. You can fine-tune the results by manually tweaking the position of specific sliders. In this image, dragging Orange up to +33 lightens some of the subject's skin tones and makes them stand out against the shadows.

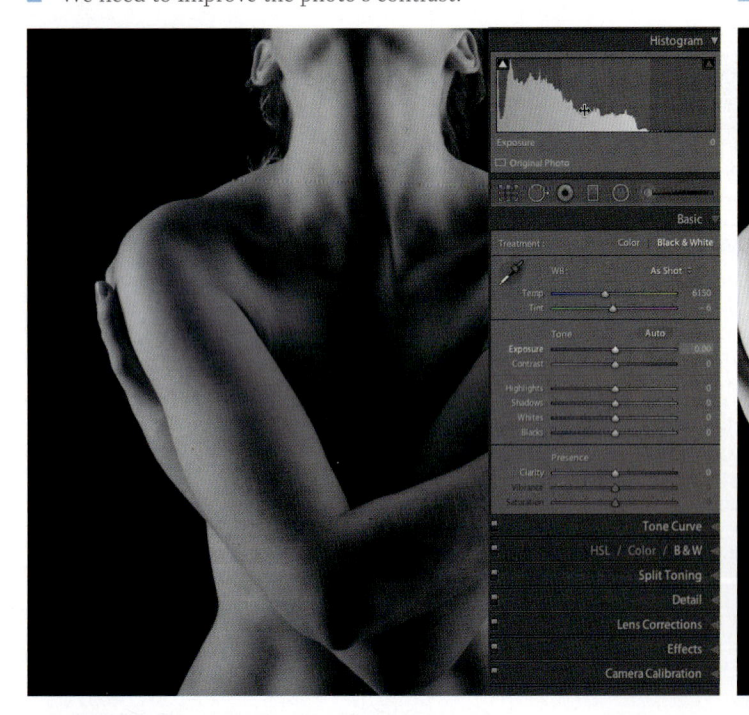

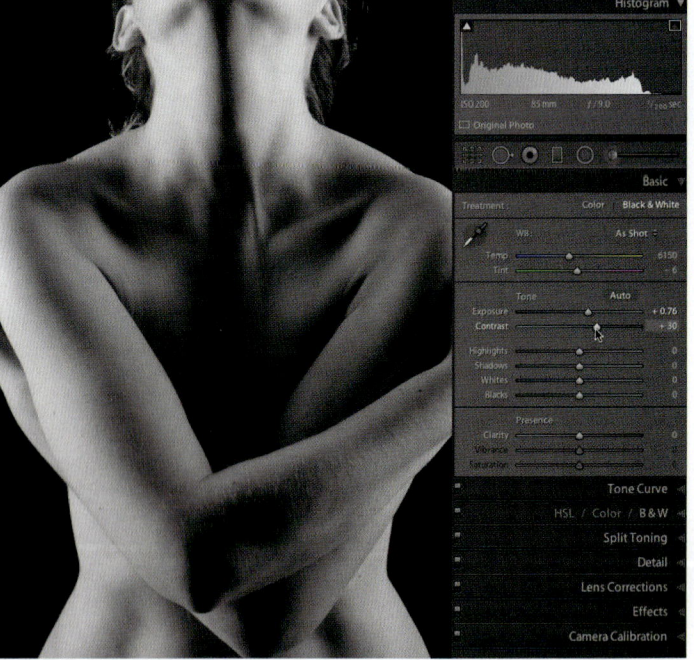

3 Examine the histogram
After lightening greyscale tones that correspond to orange colours in the starting image, the tones in the histogram window's graph have slid right towards the highlight section. We still need to boost the contrast, because an effective black-and-white photo should have some black shadows and bright white highlights. If you move the cursor over the middle of the graph you'll see that the Exposure slider influences this section.

4 Increase the exposure
Click the middle of the histogram window and drag right to push the Exposure up to +0.76 (or drag the Exposure slider to the right if you prefer). Now that you've remapped the photo's weaker midtones to a brighter tonal level, this gives the graph a wider spread of tones. Increase the strength of the tones by pushing the Contrast slider up to +28. The higher contrast photo now has more impact, and you have a healthier-looking histogram.

Teach yourself Lightroom

SPECIAL EFFECTS

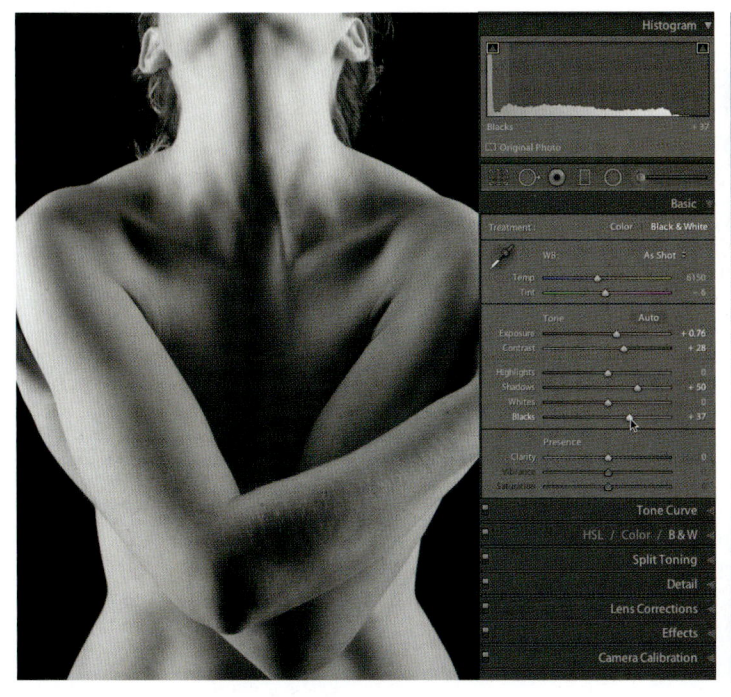

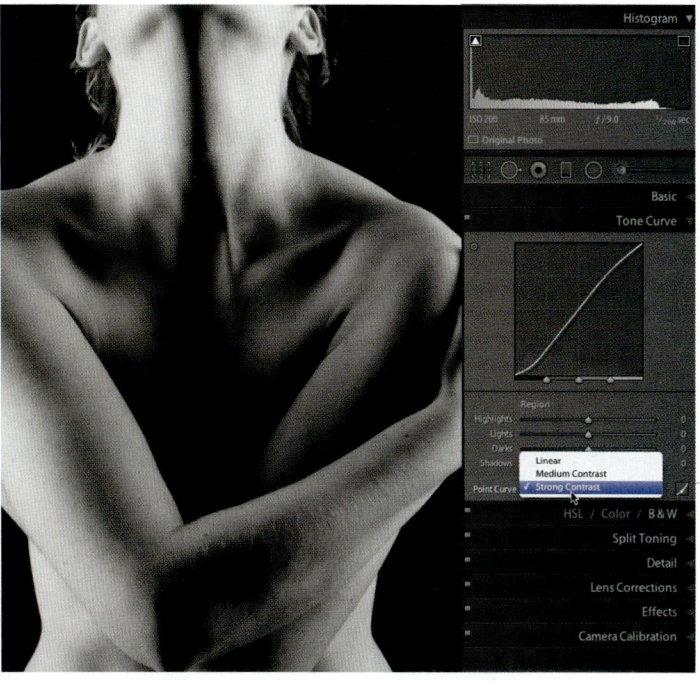

5 Lighten the shadows
To check that you can see darker detail, click the Shadow Clipping Warning icon at the top left of the histogram window. Areas that will print as pure black will appear as patches of blue. In this case, none of the model's details are clipped, but some the shadows lack detail. Push the Shadows slider up to +50. Reveal more detail in the model's darkest contours by dragging the Blacks slider to +37.

6 Boost the contrast
To help reveal the shape and form of our fine-art nude we can push the contrast even further, courtesy of the Tone Curve tab. This powerful tool enables you to selectively lighten or darken tones in a variety of ways, as you'll see from our in-depth look at the Tone Curve in the following chapter. In this instance, simply click the Point Curve drop-down menu and choose Strong Contrast.

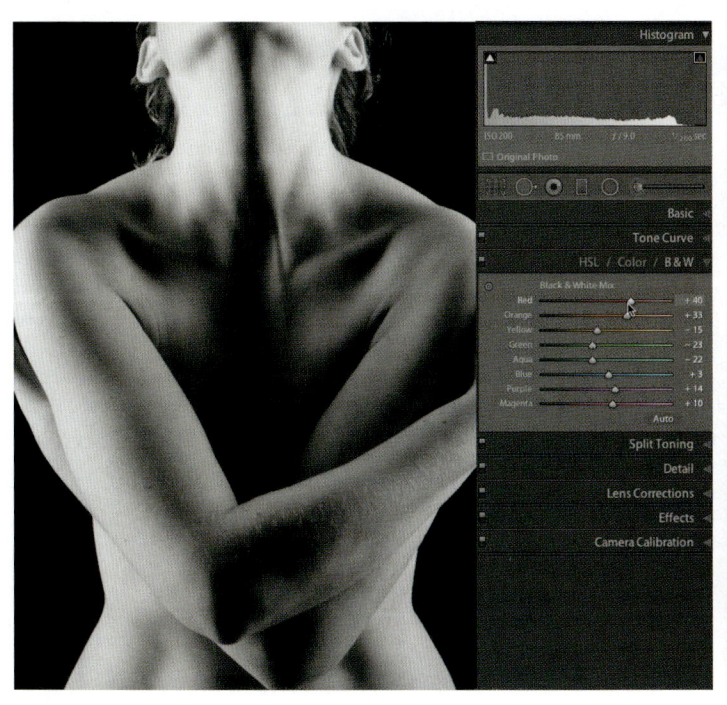

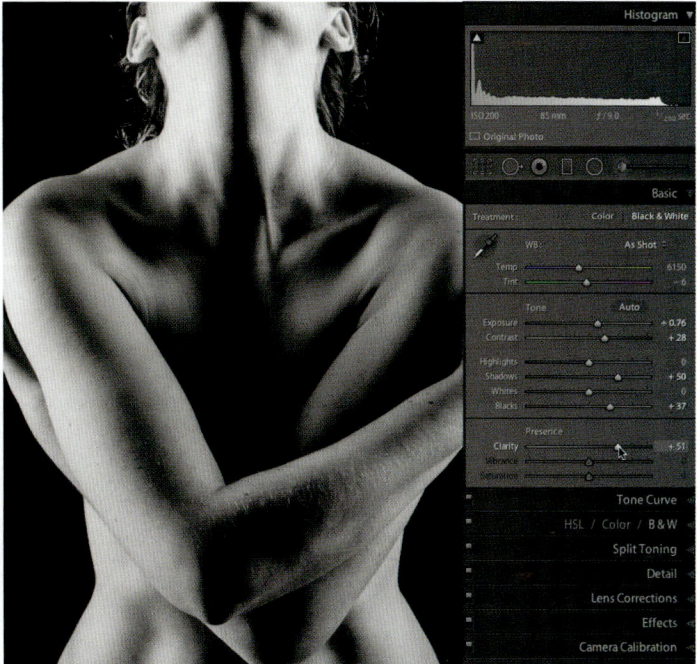

7 Fine-tune the greys
This contrast boost reintroduces some clipping in the background shadows, but there are no important details in these regions to worry about. Turn the Shadow Clipping Warning off. You can now go back to the B&W tab and fine-tune the conversion. Push Reds up to +40 to lighten the greys that correspond to this range of colour in the original photo. We now have a striking high-contrast mono conversion.

8 Add the finishing touches
Some of the model's curves are represented by subtle greyscale tones. To help emphasise these areas, go back to the Basic panel and go to the Presence section. Drag the Clarity slider right to a value of +51. This increase in Clarity boosts the contrast of the fine midtones to help emphasise the subject's curves and contours. It also gives the finished photograph a bit more punch.

Teach yourself Lightroom

SPECIAL EFFECTS

GET THE FILES HERE: http://bit.ly/tylr2016

Create HDR images

Discover how to combine several exposures into a single, highly detailed image using Lightroom's HDR Merge feature

Lightroom's HDR feature enables you to merge several different exposures into one, providing a high dynamic range with detail in the brightest highlights and darkest shadows (which makes the feature particularly useful for high-contrast landscapes and interiors). Best of all, it then creates a merged DNG raw file packed with tonal information. Understandably, the feature isn't as in-depth as dedicated HDR software like Photomatix or Nik's HDR Efex Pro, or even Photoshop's own Merge to HDR command. But Lightroom's approach is slightly different. It creates a detail-rich raw file without resulting in the over-processed look that puts many photographers off HDR. As such, it's more HDR as a utility than as a style, and all the better for it. While limited, the options in the Merge HDR command are very effective. There's an Auto Align feature, so as long as camera movement isn't too severe, you can get away with merging hand-held sequences. There's also a Deghost control to fix movement within the frame.

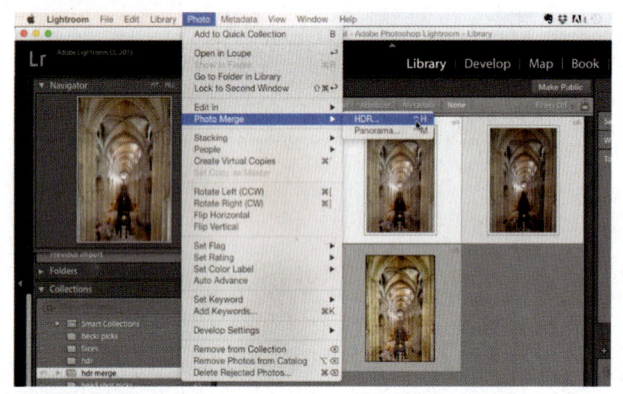

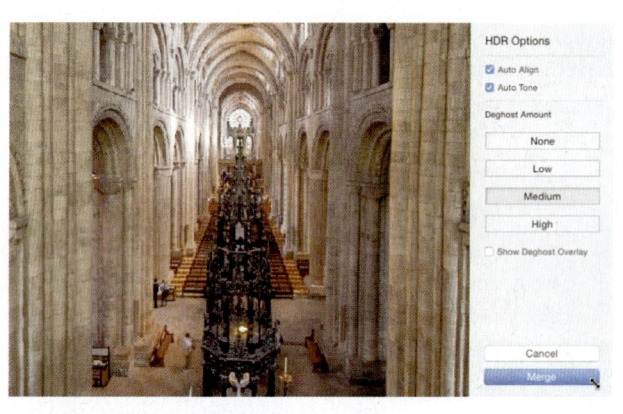

1 Start HDR Merge
You'll need a set of images taken in alignment, with each frame set at a different exposure. Once the images are imported into Lightroom, Shift-click between the first and last frame to select them. There are three ways to begin the HDR command: go to Photo>Photo Merge>HDR, or right-click the images and choose Photomerge, or simply press Cmd/Ctrl+H.

2 Improve the tones
The Auto Align box to the top-right will help if the alignment between each frame is slightly off. The Auto Tone box will bring out detail in shadows and highlights. Bear in mind that this is non-destructive, so once you merge the images, you can change the Auto Tone settings with the Develop module's Basic panel.

3 Fix ghosting movement
The Deghost amount will help to correct movement within the frame — perhaps from moving grass or clouds, or in this case people — by taking the problematic area from a single frame rather than by merging several. Along with four deghosting strengths, there's also a helpful 'Show deghost overlay' option that enables you to see exactly which areas are being corrected. We used Medium here.

4 Merge and enhance
Click Merge and Lightroom will create a new raw DNG file with the suffix HDR and add it to your Library. Take the new image into the Develop module for further enhancements. There's far more tonal information than in a normal photo (for example, Exposure goes from -10 to +10 stops). Use the tonal tools and make selective adjustments with the Adjustment Brush to improve the image.

Teach yourself **Lightroom**
SPECIAL EFFECTS

AFTER

BEFORE

BEFORE

BEFORE

BEFORE

83

Teach yourself **Lightroom**
ADVANCED EDITING

Teach yourself **Lightroom**
ADVANCED EDITING

Advanced editing

Improve tones, reduce noise, apply presets and more with Lightroom's advanced tools

86 Introducing the Tone Curve panel
Take more precise control over the tones in your photographs using the power of Lightroom's Tone Curve panel

88 Make Tone Curve adjustments
Use the powerful parametic and targeted adjustment tools in the Tone Curve to slide your way to a better photo

90 Sharpen up your images
Give soft-looking photo subjects more impact while keeping unwanted sharpening artefacts at bay

94 Reduce noise while preserving detail
Smooth out chroma and luminance noise in your photos while still preserving important image detail

96 Make changes in the Camera Calibration panel
Make quick adjustments to colour and tone by choosing from a range of camera-processing profiles in the Camera Calibration panel

98 Lightroom's powerful editing presets
Make instant creative edits to colours and tones in photos for quick enhancements or as starting points for further editing

Teach yourself Lightroom
ADVANCED EDITING

GET THE FILES HERE: http://bit.ly/tylr2016

Introducing the Tone Curve panel

Take more precise control over the tones in your photographs using the power of Lightroom's Tone Curve Panel

When shooting a high-contrast scene it can be a challenge to capture detail throughout that scene's full tonal range. If the camera's metering mode attempts to capture detail in a bright area (such as a background window) then a shaded room may be plunged into shadow. If the camera meters to capture detail in shaded areas, then the shot's sunlit sections may blow out. When you review a photo on your camera's LCD you may not see any detail in the brighter highlights, but if you're shooting in raw then Lightroom's Develop module will come to the rescue.

Because raw files are always packed full of information about a scene's tonal range, you can overcome common exposure problems with ease, and restore detail to clipped areas. As we demonstrated in chapter four, Lightroom's Develop module has a collection of sliders that enable you to target and tweak specific tones such as under-exposed shadows, without altering correctly exposed highlights. We also used the basic panel's Clarity slider to tease out more midtone detail in the shot.

The Basic panel is useful and easy to use, but Lightroom provides an alternative and more powerful way to target and tweak tones with greater precision...

Teach yourself Lightroom
ADVANCED EDITING

Lightroom Anatomy The Tone Curve tab
Get to know the features of this powerful tone-tweaking tool

1 CYCLE VIEWS
It's always a good idea to compare an edited photo with the original, especially when tinkering with the powerful Tone Curve. Click here to cycle between different before-and-after layouts. You can see that the after version is brighter and better exposed.

2 CURVE
Initially, the curve is a straight diagonal line, so it doesn't make any changes to the photo's tones until you start to bend it. The bottom left of the curve alters the darkest shadows, the middle of the curve changes the midtones, and the top-right adjusts the highlights.

3 REGIONS
The Tone Curve is broken up into four tonal regions – Highlights, Lights, Darks and Shadows. The Highlights slider controls the brightest pixels, whereas the Lights slider adjusts a wider range of highlights and midtones.

4 RANGE PREVIEW
When you click on one of the four sliders, this light grey blob highlights which parts of the curve that particular slider is controlling. Here we can see that the Darks slider influences a wide range of shadows and midtones.

5 CLICK AND DRAG
Instead of clicking and dragging a slider, you can click on part of the tone curve itself. The appropriate slider will be highlighted. If you drag the cursor upwards in the tone curve window, then the corresponding slider will move to the right.

6 POINT CURVE
If you click here, the sliders will vanish. You can then manually click anywhere on the curve to place a control point. You then drag the point up to lighten the tones, or down to darken them. You can place multiple control points to control the tones.

Understanding... TONE CURVE PRESETS

Instead of manually dragging the Tone Curve panel's sliders to lighten or darken specific tones, you can make quicker adjustments using the Point Curve's drop-down preset menu [1].

By setting the menu to Strong Contrast [2], the bottom-left section of the curve will dip down [3] to darken the shadows. The top-right part of the curve [4] will gently lift up to lighten the highlights. This S-shaped curve is often used to improve image contrast.

You can then manually tweak the Region sliders to fine-tune the results produced by these handy Tone Curve presets.

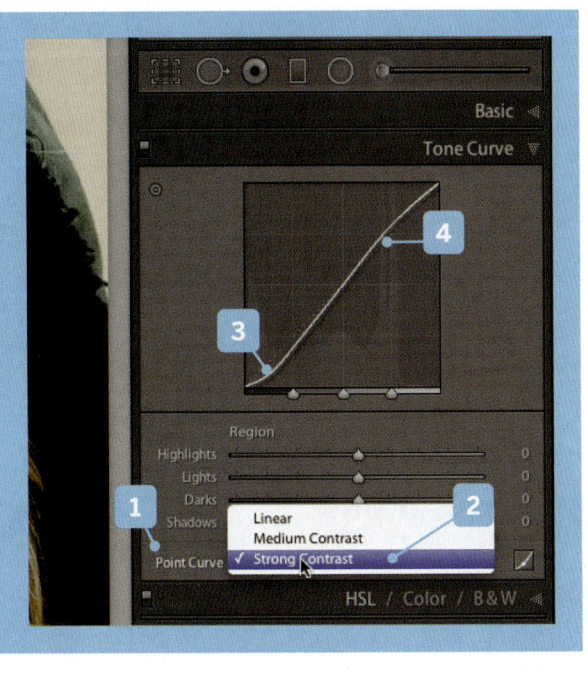

CHANGE THE SPREAD OF REGIONS

Just below the Tone Curve you'll notice three triangular-shaped handles. Between these handles are four bands of greyscale tones that indicate which tonal regions are segregated by particular handles. By default the regions are separated evenly. 0-25 covers Shadows, 25-50 covers Darks, 50-75 covers Lights and 75-100 covers Highlights. However, if you wanted the Shadows slider to influence a narrower band of shadows, drag the triangle handle left to 10. The Shadows slider will then tweak shadows, but will leave more of the midtones untouched. Click Reset to restore the region segregation handles to their default positions if you come unstuck.

Teach yourself Lightroom
ADVANCED EDITING

GET THE FILES HERE: http://bit.ly/tylr2016

Make Tone Curve adjustments

Use the powerful new parametric and targeted adjustment tools in the Tone Curve to slide your way to a better photo

The Develop module's Basic panel is designed to help you reveal a photo's missing tonal details by independently targeting and adjusting the Highlights, Shadows, Whites and Blacks. This enables you to lighten or darken problematic areas without changing correctly exposed ones. The Tone Curve panel that was featured on the previous spread provides you with an additional and effective way to target and tweak specific tones. You can combine the controls in the Basic and Tone Curve panels to overcome any tone-related problem.

Lightroom has two new powerful techniques to manipulate the Tone Curve; parametric adjustment sliders and the targeted adjustment tool. The parametric adjustment sliders are useful for choosing precise values for accurate tonal changes, as they split an image into four distinct sections: Highlights, Lights, Darks and Shadows (in descending order of brightness). The targeted adjustment tool is for those tricky parts of your image where the tone is hard to judge. It enables you to drop a pinpoint on the image to alter specific tones in an image precisely, without having to guess whether it's a 'Dark' or a 'Shadow'. Let's see how these tools affect our image.

1 Make basic adjustments
Load Architect.jpg into the Develop module. In the Basic panel, drop the Temperature slider for a cooler colour palette. The photo is lacking in colour, so set Vibrance to +30. This will boost the colours that are less saturated and bring an overall colour boost to the image without clipping them.

2 Parametric adjustments
Our highlights are too bright and shadows too dark, so to reduce the contrast in the image with lots of control, reduce Highlights by -33 and boost Shadows by +50. The darks in the image are a little bright now, so to add depth to the recesses of the model's clothes, reduce Darks by -9.

3 Targeted adjustment tool
Using the targeted adjustment tool, click and drag up on the Shadows in the folds of the overalls to lighten these areas, then click and drag down on the highlights on the model's forehead to darken the Lights and Highlights. We are left with a more natural and textured image.

Teach yourself **Lightroom**
ADVANCED EDITING

BEFORE

AFTER

Teach yourself Lightroom

ADVANCED EDITING

GET THE FILES HERE: http://bit.ly/tylr2016

Sharpen up your images

Give soft-looking photo subjects more impact while keeping sharpening artefacts at bay

Getting your photos looking pin-sharp can be a challenge for a variety of reasons. Many digital cameras have a built-in filter that blurs the photo a little in an attempt to avoid producing moiré patterns. This low-pass (or anti-aliasing) filter can also soften important details such as our flower's fine stigmas.

By placing the camera close to a small subject (such as a flower) you risk getting a shallow depth of field. This means that only a narrow band of detail will be in focus. Features in front of and behind this sharply focused zone will look blurred. An image that appears to look sharp on the camera's small LCD may turn out to look disappointingly soft when examined on your PC's larger screen.

Fortunately, Lightroom's rather aptly named Detail panel is packed full of tools that are designed to tease out the fine details in a soft-looking photo. These post-production sharpening tools work their magic by increasing the contrast around the edges of details in the image, giving them more impact. However, when you digitally sharpen a photo using Lightroom's sliders you risk exaggerating picture noise in smooth areas such as the clear blue background in our starting image. You can also introduce artefacts such as blown-out highlights, clipped shadows and distracting haloes to the sharpened areas.

In the following walkthrough we'll examine ways to sharpen your photos while keeping artefacts at bay. We'll demonstrate how the sophisticated Detail panel enables you to restrict the sharpening to important areas, while protecting other sections from being sharpened — and therefore minimising artefacts.

BEFORE

Teach yourself **Lightroom**
ADVANCED EDITING

AFTER

Teach yourself Lightroom
ADVANCED EDITING

1 Zoom in
Open our TYLR36.dng starting image in Lightroom's Develop module. To accurately assess how in-focus a photo is, you'll need to look at it at 100%. Go to the Navigator panel and click on the 1:1 option at the top. You can then drag Navigator's white preview box around to examine sections of the photo and check the focus. At this magnification the flower's subtle textures and fine details look slightly soft.

2 Check the detail
Toggle open the Detail panel. This features a detail zoom window that displays sections of the photo at 100% so that you can see how sharp they really are. Click on the little crosshair icon at the top-left of the Detail panel and move the cursor over the image in the main window. Click to render a 100% size view in the Detail zoom window. You can use this tool to check the sharpness of a photo that's zoomed out in the main window.

3 Experiment with the Amount
Lightroom will apply a sharpening amount of 25 to all photos to counteract the blurring effect of the camera's low-pass filter. If you reduce Amount to 0 you'll see that the unedited photo looks softer, so it certainly needs some post-production sharpening. All the other sliders will be greyed out. Before we sharpen the photo properly, we'll over-sharpen it to demonstrate the type of ugly artefacts that you'll need to avoid.

4 Recognise artefacts
Drag the Amount slider to 150 to increase the contrast around the edges of small details in the photo. This maximum setting makes the photo look sharper, but it exaggerates picture noise. The Radius slider increases the spread of the contrast change produced by the Amount slider. To see how it works, drag it up to 3. Now you can see ugly artefacts such as white and dark haloes clinging to the edges of the petals.

Teach yourself Lightroom
ADVANCED EDITING

5 Compare before and after
Drop the Amount slider to a more subtle value of 113 to reduce the strength of the visible ugly haloes. Drop Radius to 2.0 to decrease the spread of the edge contrast change and reduce the haloes even more. To see how your sharpened version compares with the original, toggle the Detail panel on and off. Alternatively, click the Before and After icon at the bottom-left of the workspace.

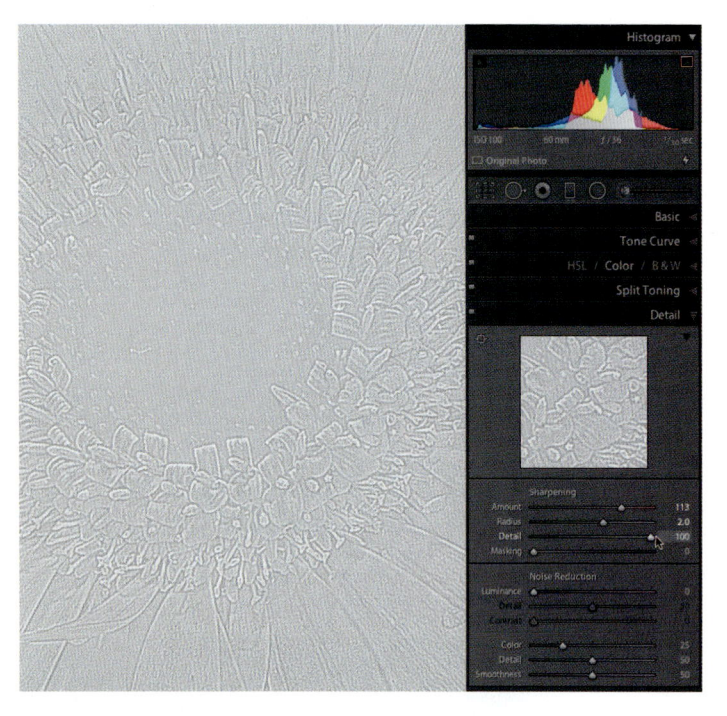

6 Balance detail versus noise
The Amount and Radius sliders have given the details in the image more impact, but they've also exaggerated picture noise. This noise is noticeable in the smooth areas such as the blue background. The Detail tab is designed to get a balance between sharpening detail while keeping noise at bay. Hold down Alt and drag Detail to 100%. A greyscale preview shows you sharp detail plus noticeable noise.

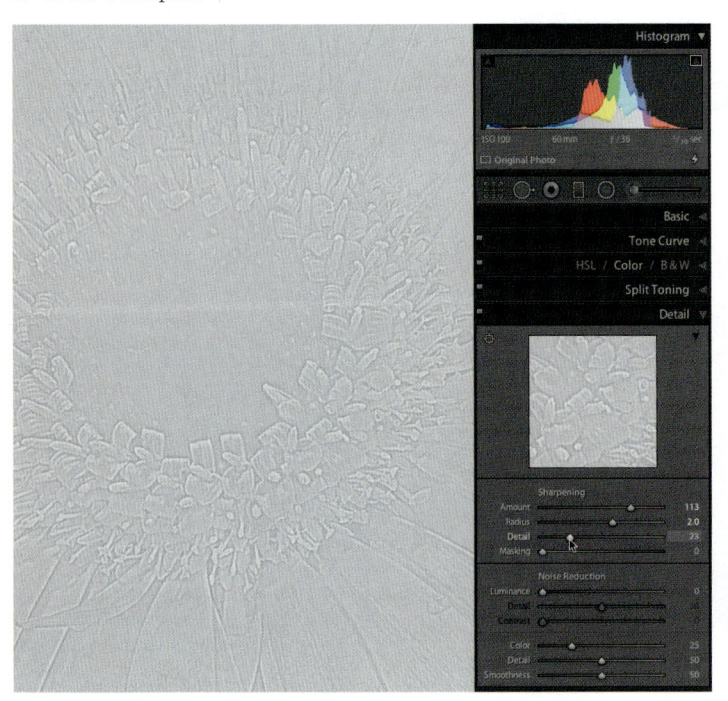

7 Reduce the noise
Hold down Alt and drag Detail down to 23. This lower setting reduces the presence of the noise in the sharpened areas, but it still adds impact and definition to important details such as the petal edges. Lightroom's Detail tab also has a Masking slider. This works in tandem with the Details slider to restrict the sharpening contrast change to important areas while keeping noise at bay. It pays to experiment with this slider with most images.

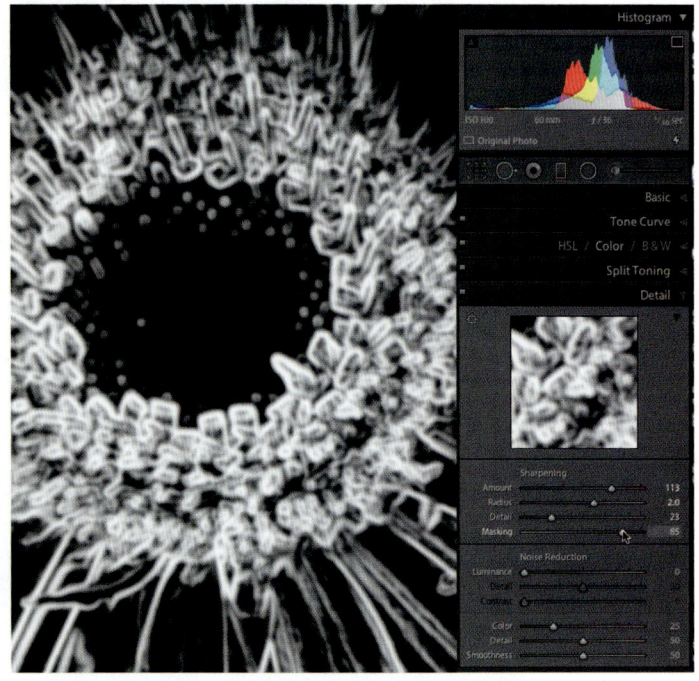

8 Mask out the smooth areas
Hold down Alt and click the Masking slider. At a value of 0, the screen will turn white. This indicates that no masking is occurring. The other three sharpening sliders are free to alter the entire image. As you hold down Alt and click and drag the slider right, the black (masked) areas appear. These masks protect parts of the photograph from being sharpened. Only the white areas are sharpened.

Teach yourself Lightroom

ADVANCED EDITING

GET THE FILES HERE: http://bit.ly/tylr2016

Reduce noise

Smooth out chroma and luminance noise while preserving important image detail

On the previous spread we demonstrated how the Detail panel enables you to sharpen a photo while keeping noise at bay. Noise is added to a photo when using a high ISO sensitivity setting on your camera, or when shooting at night with a slow shutter speed. There are two types of noise — luminance (greyscale) and chroma (or colour). Luminance noise appears throughout the image's tonal range as tiny dots. Chroma noise manifests itself as tiny patches of distracting rainbow colours, which are especially noticeable in detail-free sections of a JPEG image, such as a sky.

Luminance noise can be compared to character-adding film grain, so it isn't as distracting or undesirable as chroma noise. We'll show you how to use the Detail panel's Noise Reduction tools to get a balance between smoothing out luminance noise while preserving important image detail. It's much easier to remove chroma noise without losing detail, as you'll see in our walkthrough.

1 Improve the exposure
Open TYLR35.dng in the Library module. Toggle open the Metadata tab. We used a noise-inducing ISO Speed Rating of 12800. Zoom in to see the noise more clearly. In the Develop module, drag Exposure to +1.75, set Blacks to -64 and Vibrance to +29.

2 Reduce the colour noise
In the Detail panel's Noise Reduction section, drag Color to 0 to see the speckles of chroma noise. Slide it back to 25 to remove them. The Detail tab provides a balance between reducing colour bleed and colour speckling. Drag it right to reduce bleed or left to reduce speckling.

3 Reduce the luminance noise
The Smoothness slider is designed to remove low frequency colour mottling, but in this case the default setting of 50 produces good results. To soften the remaining luminance noise, drag the Luminance slider to 32. This reduces the luminance noise, but it also blurs image detail.

4 Preserve the detail
The Detail tab enables you to get a balance between smoothing out luminance noise while protecting detail. Drag it right to sharpen blurred details, or left to blur noise. The default of 50 produces an effective compromise. You could also try increasing Contrast.

Teach yourself **Lightroom**
ADVANCED EDITING

AFTER

BEFORE

Teach yourself Lightroom
ADVANCED EDITING

GET THE FILES HERE: http://bit.ly/tylr2016

Camera Calibration

Make quick adjustments to colour and tone by choosing from a range of camera-processing profiles

Many digital cameras provide the opportunity to process a photograph immediately as they capture it, using a range of built-in presets such as Portrait, Landscape, Neutral and so on. These presets change the look of the captured photos' colours and tones. For example, a camera's Landscape preset might boost the saturation of blues and greens. The Neutral preset will avoid boosting the colour and contrast so that you can get the look you want in Lightroom.

If you shoot in JPEG format, then the results of these in-camera presets will be harder to alter in Lightroom. However, if you shoot in your camera's raw format then you can experiment with different looks quickly and effectively in Lightroom, courtesy of the profiles and sliders in the Camera Calibration panel. This panel provides you with a springboard towards adjusting colour and tone. You can then fine-tune the results of a particular preset profile using the sliders in the Basic panel.

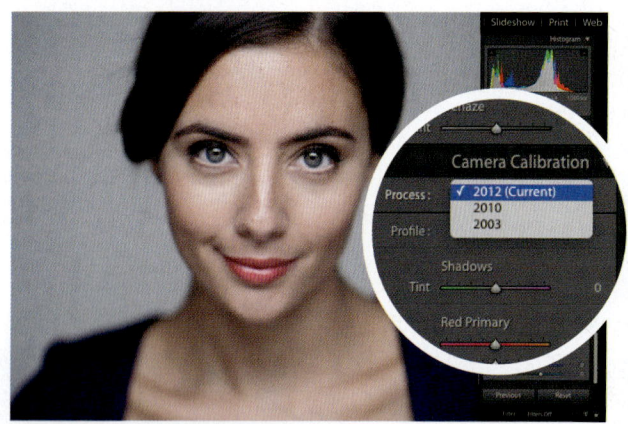

1 Process options
Open Portrait.dng in Lightroom's Develop module. Open the Camera Calibration panel. Lightroom 5 uses the 2012 raw processing engine, so jump back to the 2010 process if you prefer to replace the Basic panel's Highlights and Shadows slider with the Recovery and Fill Light sliders.

2 Choose a profile
We'll leave Process set to 2012 (Current) so that we're using Lightroom's latest raw-processing tools. Camera Landscape makes the skin look too orange. Camera Portrait increases contrast without boosting colour, so we'll settle with that profile as a starting point.

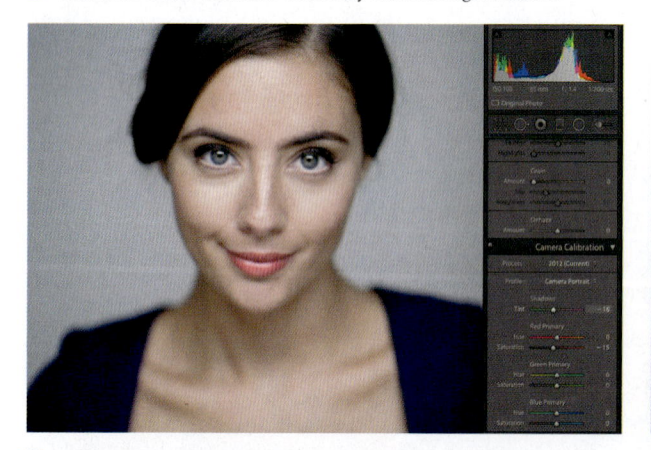

3 Fine-tune the profile
To counteract the slight magenta tint, drag the Shadows slider to the left (-16). To make the skin look less orange, go to Red Primary and drag Saturation down to the left (-15). If you're unhappy with the look, hold down Alt and click a label such as Reset Shadows.

4 Back to Basic
Finish off in the Basic panel. The shadows are a little dark so boost them to +45 to regain detail. The image may look a little washed out, so increase the contrast to +20. Now increase the Vibrance to +36 for a subtle colour boost. The Camera Calibration panel is an effective tool to try.

Teach yourself Lightroom
ADVANCED EDITING

AFTER

BEFORE

97

Teach yourself Lightroom
ADVANCED EDITING

GET THE FILES HERE: http://bit.ly/tylr2016

Powerful presets

Make instant creative edits to colours and tones in photos

One way to discover more about how Lightroom's sliders and panels function is to experiment with presets. In the previous chapter we demonstrated how you could use presets to change a video clip's colours and tones using the Library module's Quick Develop panel. However, video clips only have access to a limited range of preset effects. When editing raw files you can take them into the more sophisticated Develop module and apply a wider range of effects to your photos, as you'll see on this spread. You can use presets as a starting point for a range of looks and then fine-tune them to customise the results.

BEFORE

1 Preview the presets
Open TYLR11.dng in the Develop module. Toggle open the Presets panel below the Navigator. Toggle open a folder such as Lightroom B&W Filter Presets. As you move the cursor over each preset you'll see a preview of the look that it will produce in the Navigator window.

2 Apply a preset
To apply a preset to the photo, simply click a preset such as Blue Filter. This instantly converts the photo to monochrome and adjusts the sliders in the B&W panel to dramatically darken the reds of the dress. To lighten the skin, drag Orange up to +17.

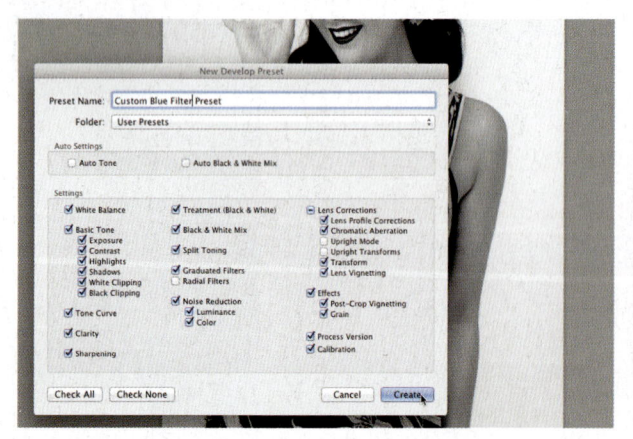

3 Save a custom preset
Once you've customised a preset effect, you can save it and apply it to other photos. Click the + icon on the right of the Presets panel. In the New Develop Preset window, label the preset. You can remove specific attributes from the preset if desired. Click Create.

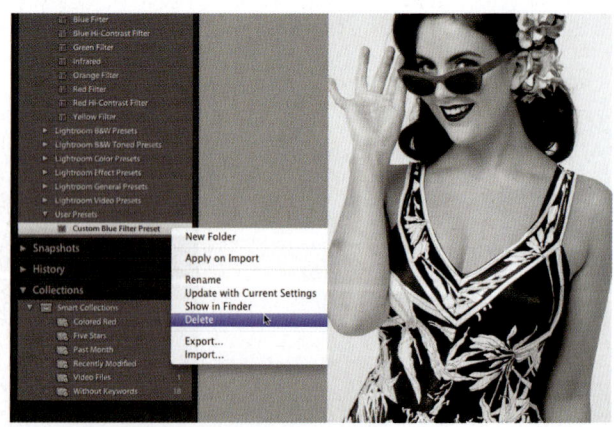

4 Open your presets
Any custom presets you create will appear in the User Preset section of the Presets window. To remove any custom preset, right-click it and choose Delete from the pop-up menu. You can also make adjustments from scratch and save them as custom User Presets.

Teach yourself **Lightroom**
ADVANCED EDITING

Take it further...

Produce a wide range of looks by experimenting with the options in Lightroom's Presets panel

Custom Blue Filter preset
Here's the result produced by the customised Blue Filter preset in our four-step walkthrough. We darkened the reds to darken the dress so that it stands out in contrast with the white background. We also lightened the girl's skin for a less grey look.

Yellow Filter preset
This preset filter applies values that dramatically lighten the reds of the dress and the oranges of the subject's skin to produce a smoother milky complexion. This helps her eyes and hair to stand out from the other elements in the photo.

Cross Process 2 preset
This Lightroom preset mimics a traditional darkroom processing technique by tweaking the Tone Curve and adjusting the HSL panel's Saturation and Luminance sliders. We'll explore cross-processing in more detail in the following chapter.

Vignette 2 preset
This preset can be found in the Lightroom Effect Presets folder. It leaves the photo's colours and tones as they are, but adds vignetted corners and film-like grain. Presets are great for giving your digital photos an instant retro look.

Teach yourself **Lightroom**
GET CREATIVE

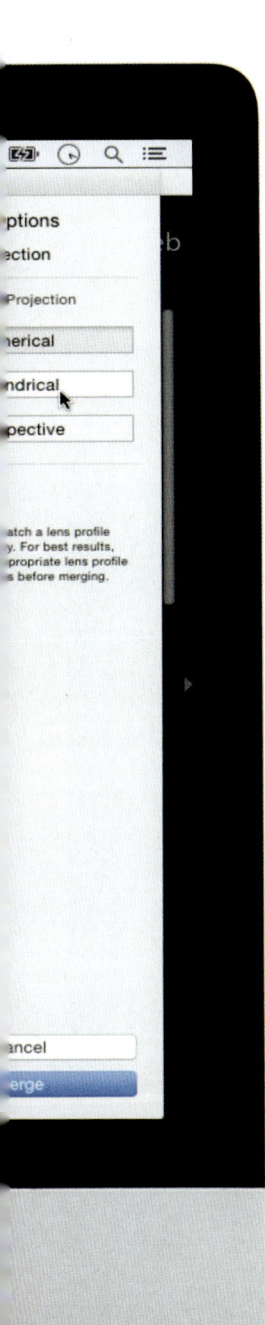

Teach yourself **Lightroom**
GET CREATIVE

Get creative

Give your images an edge by applying creative special effects in Lightroom

102 **Bring your landscapes to life in Lightroom**
Use selective adjustments and the Clarity slider to add punch and contrast to cloudy seascapes in Lightroom

104 **Try creative cross-processing**
Produce selective shifts in colour by mimicking a classic darkroom-processing technique in Lightroom

106 **Merge panoramas**
Use Lightroom's automated image-merging tool to stitch several frames into a stunning panoramic image

110 **Use Merge to HDR on your landscapes**
Make use of Lightroom's HDR Merge options to effectively combine a series of landscape photos into one beautiful image

Teach yourself **Lightroom**
GET CREATIVE

GET THE FILES HERE: http://bit.ly/tylr2016

Bring your landscapes to life
Use selective adjustments and Clarity to add punch and contrast to cloudy seascapes in Lightroom

BEFORE

1 Tweak the colour
Before adjusting individual areas of the image, make some basic adjustments to the whole image. The main adjustments are to Vibrance and Saturation. Set Vibrance to around -60 and Saturation to around +60 to boost the blues and greens to enhance the cool look of a winter landscape.

2 Enhance the land
To bring back some detail in the headland, select the Adjustment brush and paint over this area. If you paint over the sky or sea, use the Erase option to correct it. Drag Exposure to around +3, Shadows to +64 and Contrast to 75. Increase Clarity and Saturation to around +30.

Teach yourself Lightroom
GET CREATIVE

AFTER

3 Adjust the sky
Select a new Adjustment brush and paint over the entire sky. Add more contrast to the sky by setting Clarity to 75 and Contrast to 95. To remove some of the unwanted blue colouring set Saturation to -50. Finally, set Highlights to -76 to retain detail in the sky.

4 Improve the sea
The final area to adjust is the sea. Pick the Graduated Filter tool and click and drag the filter from the bottom of the image upwards until the centre line is aligned with the horizon in the image. Set Temperature to 17, Contrast to 20, Highlights to 12, Shadows to -39 and Clarity to 71.

Teach yourself **Lightroom**

GET CREATIVE

GET THE FILES HERE: http://bit.ly/tylr2016

Try creative cross-processing

Produce selective shifts in colour by mimicking a classic darkroom processing technique

The term cross-processing refers to a developing technique used in pre-digital chemical darkrooms. It involved deliberately developing print negatives using chemicals that were designed for use with slide film (or visa versa). The use of these incorrect chemicals resulted in a shift in colour and an increase in contrast. This darkroom technique created great stylised and eye-catching images. Cross-processed blues often took on a green hue for example, while the shadows might feature a hint of magenta.

This editing process is still popular now, especially with fine art, fashion and stock photographers. In the days of the traditional darkroom the results could be a bit hit or miss, so you had to experiment to get the desired shifts in colour and tone. But Lightroom's Color panel enables you to tweak and adjust individual colour channels to replicate almost any chemical combination you might desire. We'll show you this, plus how to use the Graduated Filter to tease out more detail. This creative process elevates a standard photo to a more interesting level.

BEFORE

1 Reveal colour and detail
Import TYLR23.dng into the Develop module. In the Basic panel, set Temperature to 6050 for a cooler look. Set the Highlights slider to -96 to claw back brighter details. Drag the Shadows slider right to +69 to reveal more colour and detail in the under-exposed hull of the boat.

2 Adjust the Presence
In the Presence section, drag Clarity to +56 to make the pebbles pop out. Drop Vibrance to -19 for a subtler wash of colour. The sky looks bland, so click the Graduated Filter tool icon. Set Exposure to -1.61. Boost Contrast to +18. Set Clarity to +23 to tease out fine detail in the clouds.

Teach yourself **Lightroom**
GET CREATIVE

AFTER

3 Draw a gradient
Drag the cursor down to draw a gradient that overlaps the sky and the top section of the boat's cabin. This gradated tonal adjustment will gently darken the sky and increase its contrast. Tease out more colour information by dragging the Saturation slider to 72. Press M.

4 Cross-process the colours
In the Color panel click the Blue channel. Set the Blue channel's Hue to -48 to give the blues the classic green tint associated with cross-processing. Drop the Saturation to a more subtle -50. Click the Reds and set Hue to +54 to add a hint of orange to the red paintwork of the boat. ∎

Teach yourself Lightroom

GET CREATIVE

GET THE FILES HERE: http://bit.ly/tylr2016

Create a panorama

Use Lightroom's Merge Panorama command to combine several frames

Lightroom's Merge Panorama command stitches several horizontal or vertical frames together to create a panoramic raw file — perfect for those times when your lens can't fit everything in, or if you want to pack in extra detail.

When shooting the frames for a panorama, use a tripod to keep the camera position fixed and make sure that the panning motion remains perfectly level by checking the horizon as you pan (a spirit level comes in handy here). Shoot with your camera in vertical orientation to record the maximum amount of detail, and allow for a generous overlap between each segment.

Lightroom's Merge Panorama command offers three Projection modes that stitch the frames in different ways. Spherical maps the frames as if on the inside of a sphere. It's ideal for very wide panoramas, or ones that have several rows to them. Perspective maps the segments as if they were on a flat surface, keeping lines straight. As such, it's good for architectural or city scenes. But it can lead to extreme distortion and warping at the edges, so check them before applying. Cylindrical maps the frames as if they are on the inside of a cylinder. It's ideal for wide panoramas because distortion is minimal, and vertical lines stay straight.

BEFORE

BEFORE

1 Start Merge Panorama
First, Cmd/Ctrl+click to select all the frames to stitch into your panorama, then go to the Develop module, scroll down to the Lens Correction panel, click Profile and Enable Profile Corrections. Next, to begin the merge, select Photo›Photo Merge›Panorama, or right-click the images and choose Photomerge, or simply press Cmd/Ctrl+M.

2 Choose a projection
There are three projection modes to choose from: Spherical, Cylindrical and Perspective. Each maps out the frames in a different way. Spherical places them as if on the inside of a sphere, Cylindrical as if on the inside of a cylinder, and Perspective as if placed flat. Experiment with each. We've used Cylindrical here.

Teach yourself Lightroom
GET CREATIVE

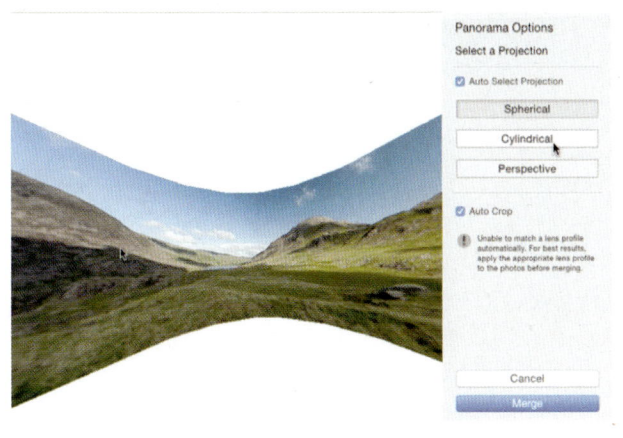

AFTER

BEFORE BEFORE BEFORE BEFORE BEFORE

3 Auto Crop messy edges
Tick the Auto Crop check box to automatically remove any messy edges to give you a tidy rectangular image. It's non-destructive and can be changed later with Lightroom's Crop tool. Try unchecking the box just to see what's being cropped off. With Perspective Projection mode here, you can see the extreme distortion at the edges.

4 Enhance the panorama
When you're happy with the settings, click Merge. The panorama will show up alongside the originals with the suffix Pano. It's a DNG raw file, so you can process it like any other raw file. Take it into the Develop module to make any changes you like. Here we've boosted the colours and added a gradient to darken the sky.

107

Teach yourself Lightroom
GET CREATIVE

GET THE FILES HERE: http://bit.ly/tylr2016

Produce detail-rich HDR panoramas

Use Lightroom's two Merge commands in combination to make a high dynamic range panorama

If you want to create detail-rich panoramic HDRs, two of Lightroom's newest features — HDR Merge and Merge Panorama — make the process quick and easy, producing images that pack in the detail while remaining natural and realistic. Shooting for HDR panoramas requires a tripod and some basic exposure control. Like you would with a normal panorama, you need to shoot the scene in overlapping segments (ideally with a vertical camera orientation). But rather than shooting a single frame for each segment, you need three frames taken at different exposure values. The easiest way to do this is to set your camera up for bracketed exposures. So for example, if your panorama has five segments, you'll end up with 15 shots.

When it comes to combining the images, first you merge the three frames for each segment into a single HDR image, then you combine all the HDRs to make a panorama. You can even set up the merging process to run in the background so you can get on with other tasks while you wait for the HDRs.

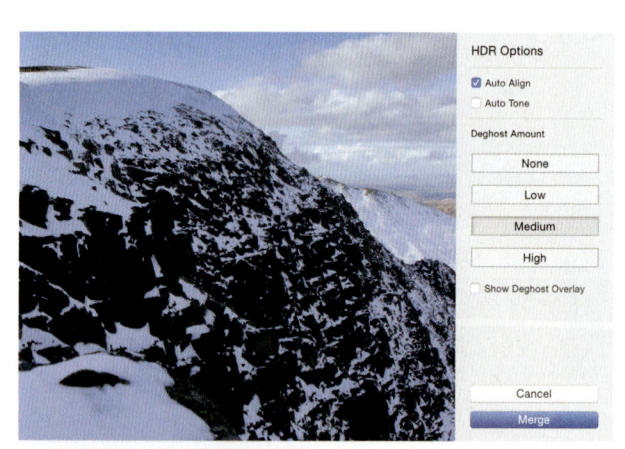

1 Merge an HDR
To create an HDR panorama you'll need a range of exposures for each segment. Cmd/Ctrl+click the three or so exposures for the first segment to select them all, then right-click them and choose Photo Merge›HDR (or press Cmd/Ctrl+H). Uncheck Auto Tone, choose a Deghost Amount to fix any movement, then click Merge.

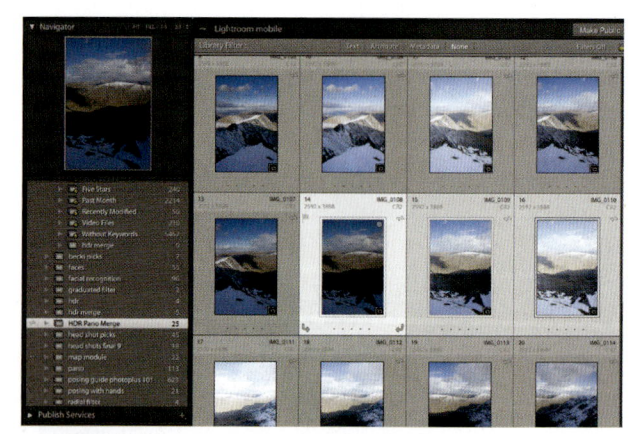

2 Run in the background
There's no need to enter the HDR settings every time. You can simply use a keyboard shortcut to run the same command in the background. Cmd/Ctrl+click to select the frames for the next segment, then press Cmd/Ctrl+Shift+H to merge (you'll see the progress bar appear at the top-left). Repeat for the rest of the segments.

Teach yourself Lightroom
GET CREATIVE

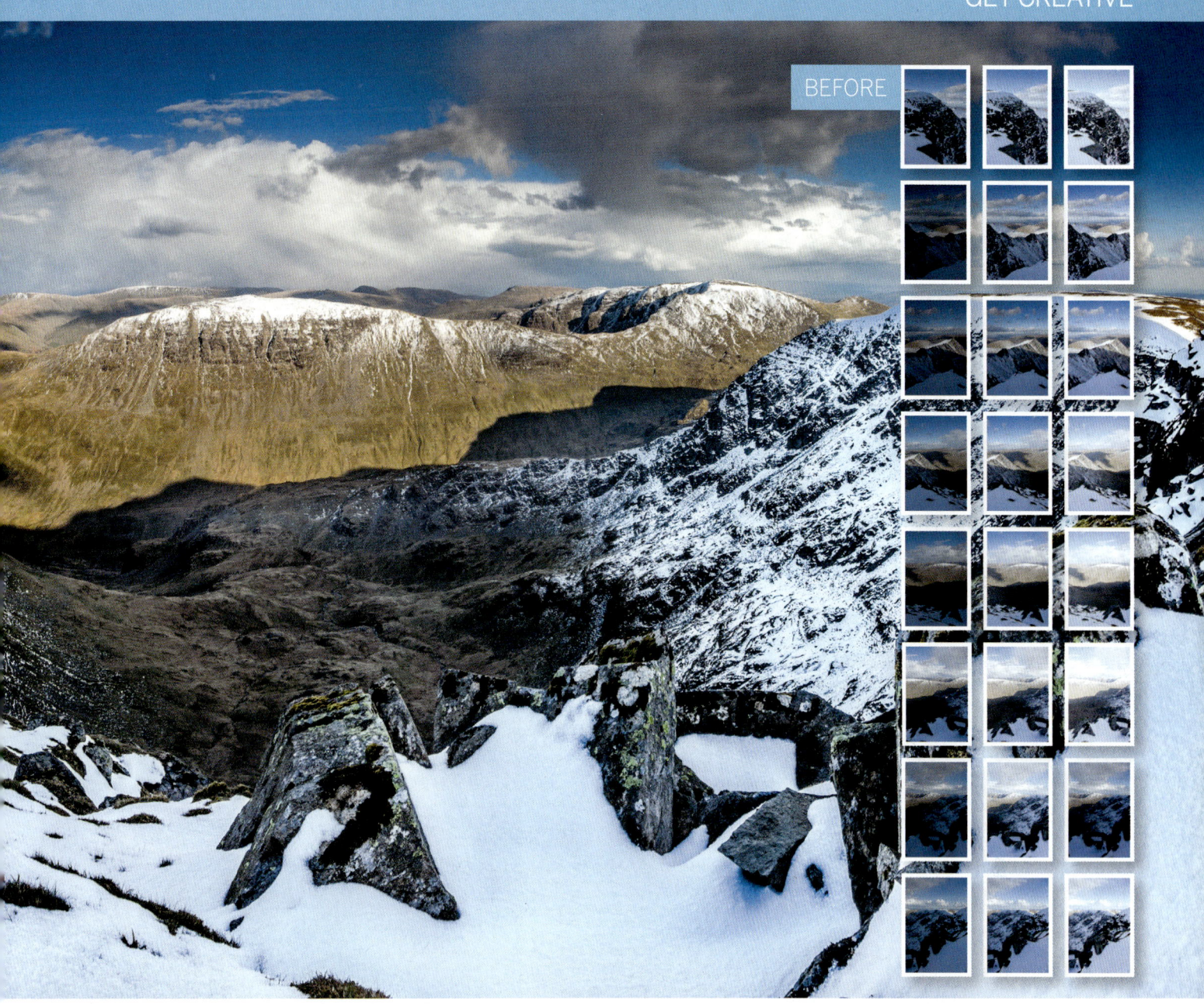

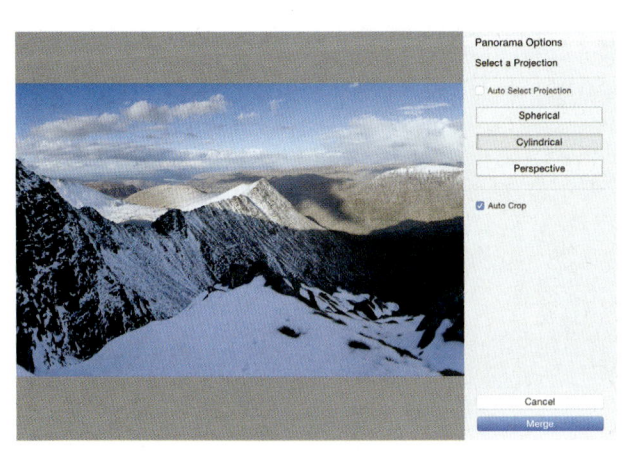

BEFORE

3 Stitch a panorama
Once all the HDRs have been created, it's time to put them together to make the panorama. First, filter out the HDRs: in Library Grid view, click Text in the Filter bar and type 'HDR'. Select all the HDRs, then press Cmd/Ctrl+M to open the Merge Panorama box. Choose a Projection (Cylindrical here), then click Merge.

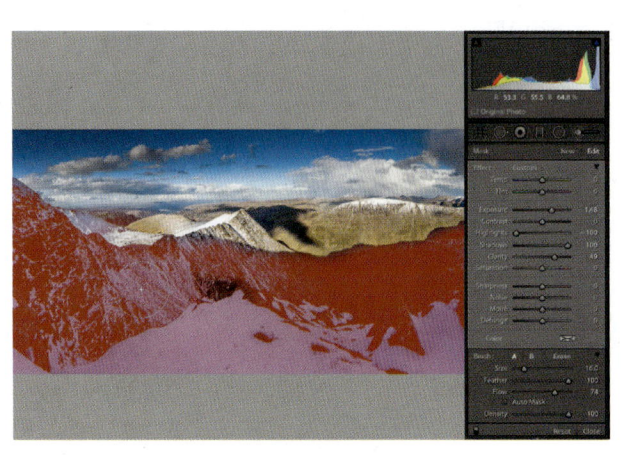

4 Tease out the details
Take the resulting panorama into the Develop module. Use the Basic panel sliders to reveal detail in the highlights and shadows, and increase Clarity to get the HDR look. Next, tease out detail with the Adjustment brush. Grab the brush from the toolbar, paint over an area, then use the sliders to change it as desired.

Teach yourself Lightroom

GET CREATIVE

Create a dramatic landscape with Merge to HDR

We walk you through how to use Lightroom's HDR Merge options to effectively combine a series of landscape photos into one beautiful image

BEFORE

AFTER

Landscape imagery is near and dear to most photographers, capturing the beauty we see as we travel through our lives. One thing you'll notice pretty quickly is that unless the sun is at your back, the sky will be much brighter than the land, especially for sunrise and sunset. Traditionally photographers would use graduated filters to help account for this, but there is a way to go without them. That way is high-dynamic-range or HDR processing.

Here you combine a series of photos at set stop intervals apart, usually two or three stops with more than two photos. (This process is called bracketing.) Most people use a combination of three photos, but five, seven or even nine are also common. Usually only the shutter speed is changed between each set.

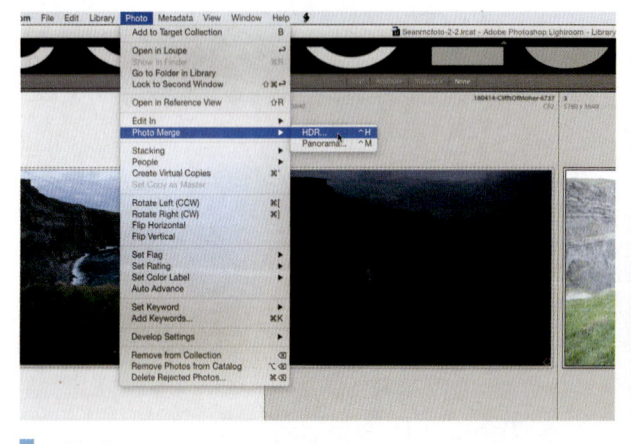

1 Select your images
Gather the photos to process together. From the Photo menu, go to Photo Merge and choose the HDR option. This set has three photos that are taken at 0EV, -3EV and +3EV. You can also use the shortcut Ctrl+H to do this (on both Mac and PC).

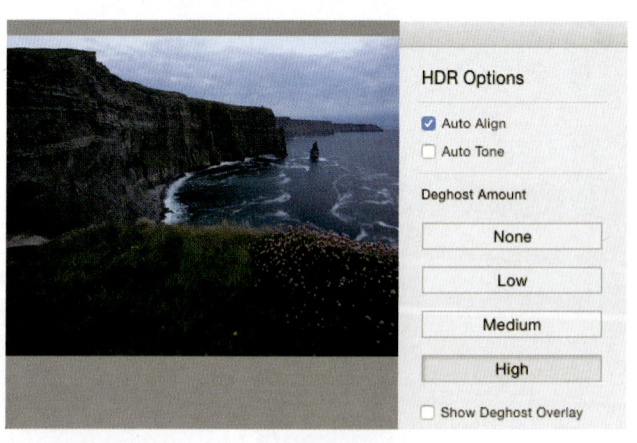

2 Auto align photos
The HDR Merge Preview dialog appears. After a short while, a preview of the HDR will appear. The top option you can select is Auto Align. This is great if you shot your images hand-held. It's even useful if you shot on a tripod, because it may have moved slightly between shots.

Teach yourself Lightroom
GET CREATIVE

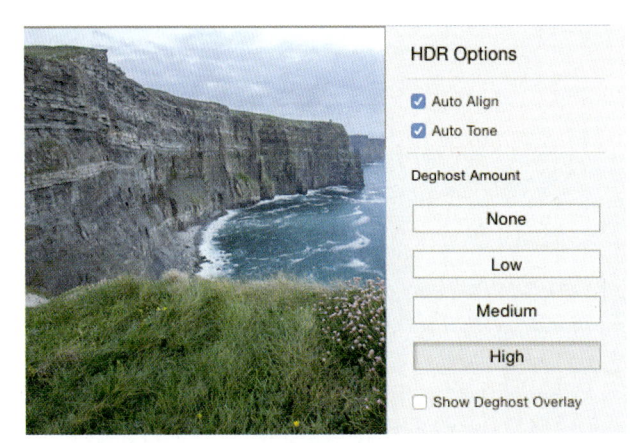

3. Apply Auto Tone
The next option is Auto Tone. This applies an auto-correction to the image, which sets the Highlights, Shadows, Whites, Exposure and Blacks of the photo. While it tends to be a little on the bright side, many photographers use this all the time and tweak from there rather than start from scratch.

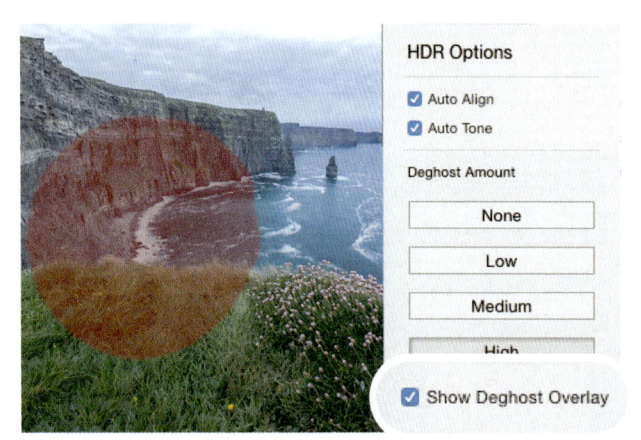

4. Remove ghosting instances
Things move between exposures: clouds, waves, people and so on. Things that appear in different places can appear as 'ghosts'. The Deghost Amount section removes these artefacts. Choose a strength. Go for High if there's been a lot of changes, or None if it looks good with it. Check what's been affected using the Show Deghost Overlay option.

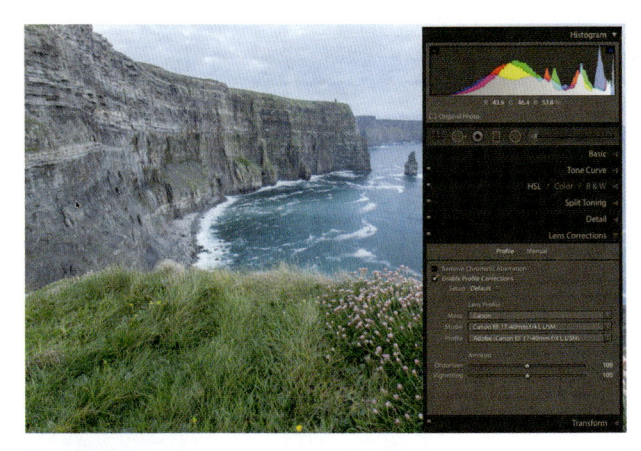

5. Produce merged HDR
Press the Merge button to create the HDR in the background. This creates the auto-toned file, which is a 32-bit DNG file with full raw control. The next step is to apply a Lens Profile to correct distortions. Rotate the file as well if needed.

6. Final tweaks
The final step is to get creative with the photo. I've worked with the Basic panel and Dehaze in the Effects panel. I set Exposure 1.50, Contrast 25, Highlights -45, Shadows +45, Whites +15, Blacks -20, Clarity, Vibrance and Saturation 35. Finally Dehaze was set to +15.

Set up right for the best results
Use a tripod for front-to-back sharpness

While Auto Align is great for hand-held HDR merging, it'll only work where your longest exposure is above the reciprocal of your focal length on full-frame. In practice, this is about 1/15 to 1/30 with a wide-angle lens and a steady hand. In order to have your photo sharp from front to back, you'll need an aperture of f/11 to f/16. f/22 can introduce too much diffraction, creating a blur that looks as bad as being out of focus. These narrow apertures require a slower shutter speed. The solution isn't just to increase the ISO: it's carrying a tripod with you and keeping your ISO low for the cleanest-looking shot.

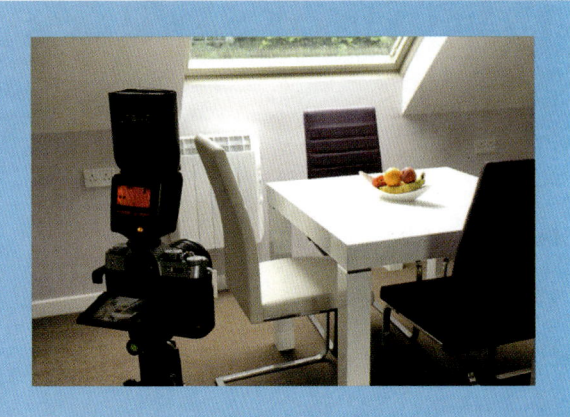

Teach yourself **Lightroom**

PRINT AND PUBLISH

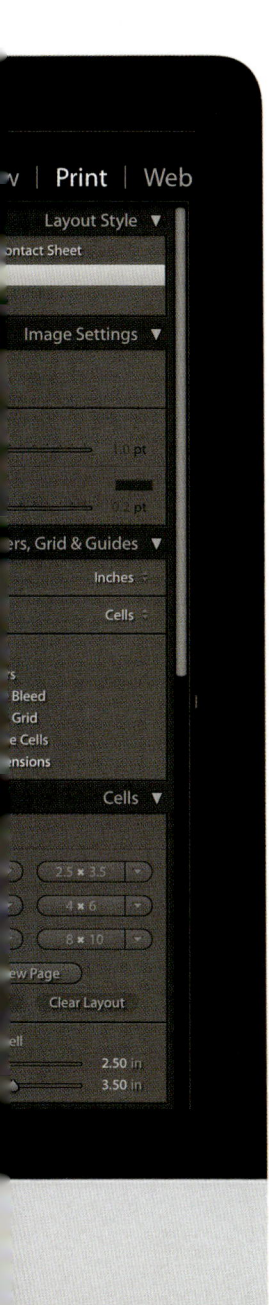

Teach yourself **Lightroom**
PRINT AND PUBLISH

Print and publish

Produce perfect prints of your photographs using Lightroom's Print module

114 Soft proof your images
Identify and correct unprintable colours so that the printed version looks similar to what you see on screen

116 Introducing the Print module
Set up your page size and orientation and use the Layout Style panel to create a contact sheet of your photographs

118 Create a custom print layout
Use the Template browser to create a range of print layouts, or create one from scratch by modifying cells

120 Watermark your images
Use the Print module's Page panel to protect your photos by adding watermarks or an Identity Plate to each image

122 Publish your photos online
Export your processed pictures to social networking sites such as Facebook, and display photos online with Lightroom Web

124 Create an online photo portfolio
Showcase your pictures in an interactive gallery using the tools and templates in Lightroom's Web module

Teach yourself Lightroom
PRINT AND PUBLISH

GET THE FILES HERE: http://bit.ly/tylr2016

Soft proof your images

Identify and correct unprintable colours so that your printed photo looks similar to what you see on the screen

In chapter five we processed the colours and tones in a landscape photograph to reveal more detail in the image and create more attractive vibrant-looking colours. After processing a photo to look good on screen, it can then be frustrating to end up with a print that looks less bright and vibrant than the digital version of the photo.

Your printer may struggle to reproduce the processed picture's colour accurately. This is because computer displays produce millions of colours by mixing reds, greens and blues (RGB) together, while most domestic printers combine cyan, magenta, black and yellow (CMYK) to create a narrower range of colours. Colours that can't be printed are referred to as 'out of gamut' colours. The image on your monitor is also brightly illuminated, leading to vibrant colours that can look comparatively drab on paper.

Lightroom's Develop module has a handy Soft Proofing mode that's designed to help you identify the out-of-gamut colours that a printer can't reproduce, so that you can adjust them to fall within the printable range. To demonstrate soft proofing, we'll use the bright and colourful processed version of our TYLR15.dng starting image that we created in the Graduated Filter tutorial in chapter five.

Lightroom uses the sRGB (standard RGB) colour space that's designed to display colours on screen. After selectively adjusting problematic colours in an sRGB colour space, we'll demonstrate how to force Lightroom to use a narrower more printer-friendly colour space — Adobe RGB (1998).

Teach yourself Lightroom

PRINT AND PUBLISH

1. Examine the colour mix
Load the processed TYLR15.dng image into the Develop module. As you move the cursor around the image you'll see that the RGB values below the histogram change. The sampled dark blue sky in our screen grab is created by a mix of 28.2% Red, 34.7% Green and 50.4% Blue, for example.

2. See a Proof Preview
To get a better idea of how the photo will look when it's printed, click the Soft Proofing button. The white backdrop represents the paper border around the print. The area below the histogram shows specific numerical values for each colour RGB colour channel, rather than a percentage.

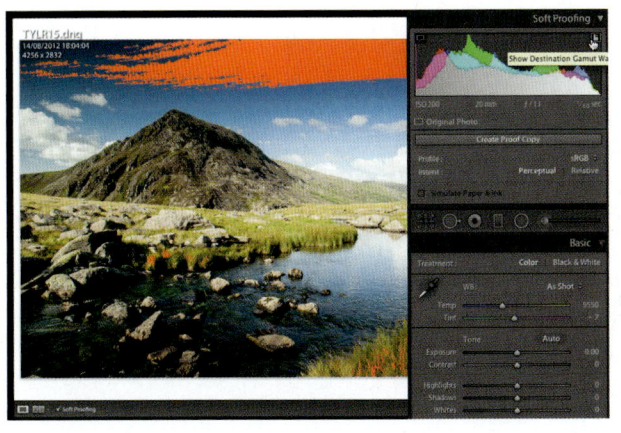

3. Activate the gamut warning
To discover which colours won't print correctly, click the Destination Gamut Warning icon at the top-right of the histogram window. Any unprintable colours will turn red. Here you can see that the darker blue pixels and some of the greens and yellows are out of gamut.

4. Create a Proof Copy
You may want a version of the image to be seen online, as well as a version that's suitable for print. Click Create Proof Copy. You'll now have two versions of the photo in the Filmstrip. To desaturate the copy's unprintable blues without changing the other colours, go to the HSL tab.

5. Desaturate select areas
Click the Saturation section of the HSL panel, and select the Targeted Adjustment tool icon. Click the sky's red patch and drag downwards to desaturate the colour. The Blue slider will slide left and the patches will vanish. Sample and desaturate the patches in the grass.

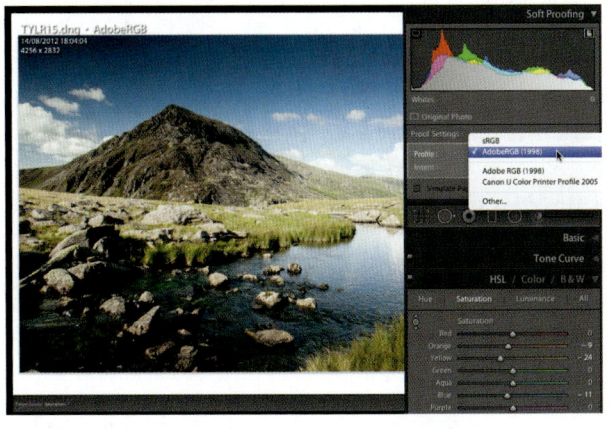

6. Change the colour profile
Alternatively, you can change the photo's colour profile menu from the default sRGB (standard RGB) to the print-friendly colour space of Adobe RGB (1998). This forces the Proof Preview's colours to conform to a printable range without the need to make selective adjustments.

Teach yourself Lightroom

PRINT AND PUBLISH

GET THE FILES HERE: http://bit.ly/tylr2016

Introducing the Print module

Set up your page size and orientation and use the Layout Style panel to create a contact sheet of your photographs

The majority of this book on using Lightroom to make the best of your digital photos has dealt with ways to organise your assets in the Library module and then process them in the Develop module. Most people will spend more time in these two modules than in the other five. However, many Lightroom users will want to share their skilfully edited photos digitally as stock photos or on social media websites and apps. We'll cover these sharing options in the last chapter.

Others will want to print their photos to place in a frame or in a physical photo album, and we'll look at this in the next few pages. On the previous spread we touched on the difference between processing a photo for on-screen display and hard copy print. Once you've used the soft-proofing tool or changed the colour profile to ensure printable colours, you're ready to take the photo (or photos) into the Print module to decide on a layout.

In the Print module you can print the photo full-size or use template layouts to create different sized prints of the same photo with a few clicks. Alternatively, you can print a range of identically sized photos as a contact sheet. The annotations on the following page will introduce you to the key tools and panels in the Print module. We'll then move on in the following pages to demonstrate how to create customised layouts that suit your specific printing requirements. So whether you want to publish on the web, via an email, or using traditional prints, we'll reveal all the key settings and tools you need to use.

Teach yourself **Lightroom**

PRINT AND PUBLISH

Lightroom Anatomy The Print module

Get to know the key features of the Print module

1 PREVIEW

In the Print module, the usual Navigator is replaced by a preview of the cells in the currently selected template. This tool helps you to understand each template's layout at a glance, and choose one that suits your printing requirements.

2 TEMPLATE BROWSER

Click here to preview and select from a wide range of layouts. You can print different sized versions of a selected photo with a range of aspect ratios. Here we have one photo in a 5x7 aspect ratio, with four images at 3.5x2.5.

3 PAGES

If you select more than one photo in the Filmstrip, each one will be placed in the template layout on a page of its own. You can click this icon to jump to the first page and use the arrows to scroll through the others.

4 SELECT PHOTOS

You can use this drop-down menu to add selected Filmstrip photos to the current template layout. Alternatively, you can choose to add Flagged Photos or even All Filmstrip Photos to a template.

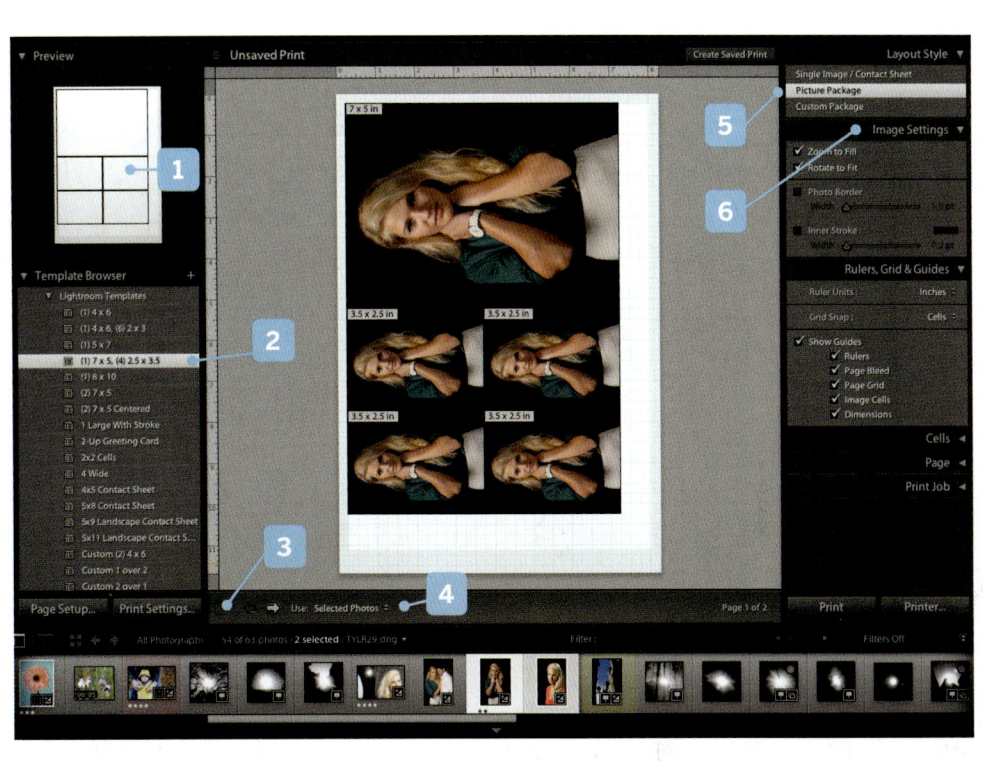

5 LAYOUT STYLE

Click this panel to print a single image or a contact sheet featuring multiple identically sized photos. An appropriate layout style will be selected as you browse through presets in the Template browser.

6 IMAGE SETTINGS

Use this panel to enlarge the image to fill the cell that it sits in. Use Rotate to Fit to make a portrait-oriented photo fill a landscape cell. You can also add borders to each cell for a classic print look.

Understanding... RULERS, GUIDES & GRIDS

Many print services still uses inches as the default unit of measurement, so you can click here [1] to display your photos using this option (or choose millimetres if you prefer).

When creating custom layouts it's worth clicking here [2] to turn on non-printable extras such as Rulers.

When creating a Picture Package of different sized photos, you can label each cell with the image dimensions by clicking here [3].

To keep images aligned, it is worth clicking here [4] and choosing Cells (or alternatively you can make them snap to the Page grid).

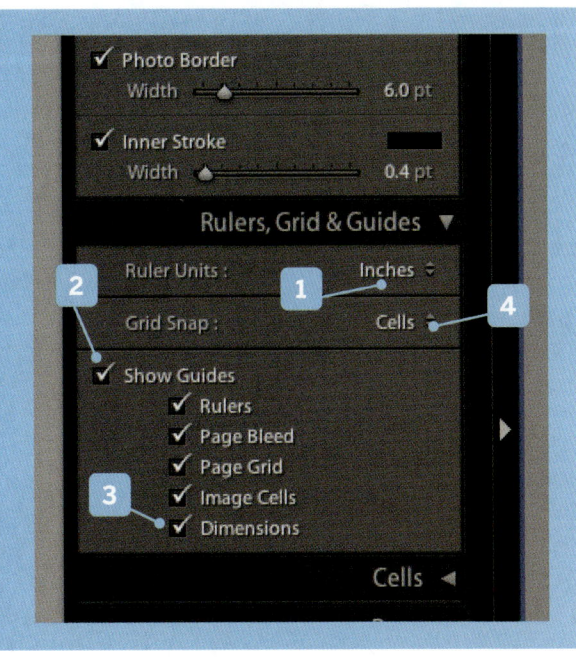

MANAGE YOUR PANELS

The Print module is packed full of panels, so you'll need to do lots of scrolling if they are all open. To keep your workspace less cluttered, right-click a panel and choose Solo Mode from the pop-up menu. You can then see one panel at a time and automatically collapse the others with a click.

With seven modules to explore, it's worth mastering the keyboard shortcuts that can summon each module with a few taps. The seven modules can be summoned with sensibly-numbered keyboard shortcuts that relate to the order in which they appear at the top of the workspace. Use Cmd/Ctrl+Alt+1 to access the Library module, Cmd/Ctrl+Alt+2 for the Develop module, and so on.

117

Teach yourself Lightroom
PRINT AND PUBLISH

GET THE FILES HERE: http://bit.ly/tylr2016

Create a custom print layout

Use the Template browser to create a range of print layouts, or create one from scratch by modifying cells

The Print module's Template browser is packed with preset templates that enable you to duplicate and resize a photo (or a selection of images) so that you can produce a Picture Package containing prints with a range of sizes. You can also print a collection of photos in the Filmstrip as a series of thumbnails, courtesy of one of Lightroom's Contact Sheet presets. In the Layout Style panel you also have the option to create a Custom Layout, which is handy if you need to print a range of different sizes that aren't available in the Template browser.

Lightroom's layout templates place your photos in cells. These cells can be manually dragged into any position on the page so you can easily change the layout of a Picture Package. You can also manually resize individual cells and set the enclosed photo to zoom to fill the cell or shrink to preserve the original composition of the image. You can also rotate cells to suit a photo's aspect ratio.

In this walkthrough we'll show you how to use the Cells panel to create a custom template from scratch and produce a unique print layout containing different-sized and shaped prints. You can save your customised layout as a new template so that you can print it at any time in the future without having to tinker with the size, shape and position of the cells in the layout.

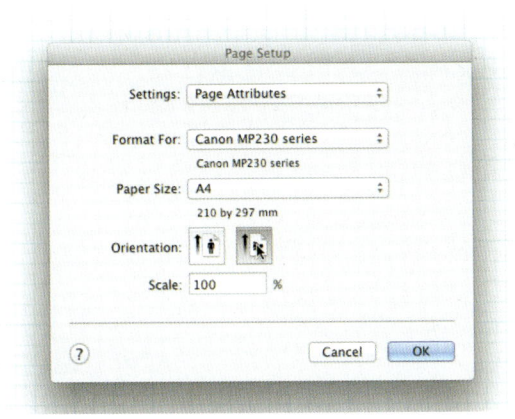

Teach yourself Lightroom

PRINT AND PUBLISH

1 Set up the page
Toggle open the Cells panel and click Clear Layout to create a blank page. Click the Page Setup button on the bottom left of the workspace. Click the landscape orientation icon and select a suitable paper size, such as A4. Leave Scale at 100%. Click OK.

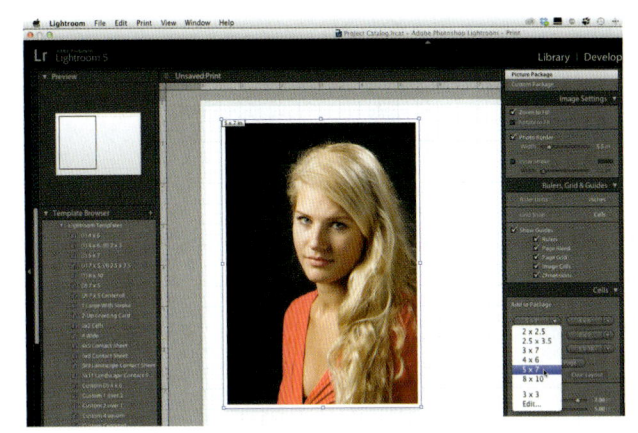

2 Add the cells
Select the TYLR27.dng starting image in the Filmstrip. The Cells panel contains icons that create cells with specific dimensions. Click the arrow to the right of each button to access other shapes and sizes. Choose 5x7 to create a large print. Tick Zoom to Fill in the Image Settings panel.

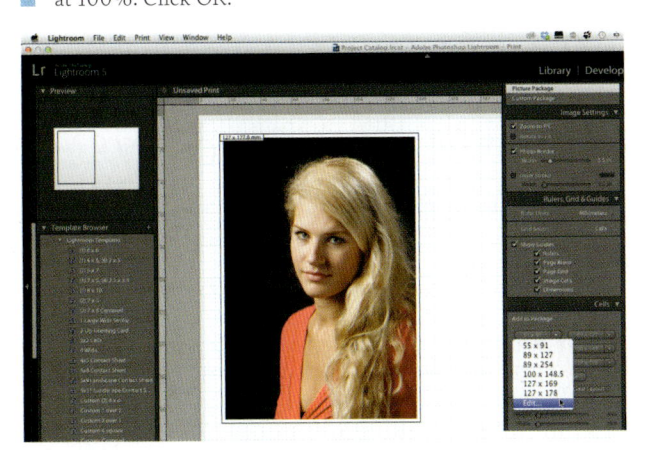

3 Change the units
Our 5x7 print will fill a standard picture frame. You may also want to print passport-sized photos on the same sheet. If so, then go to Rulers, Grid & Guides and set Ruler Units to Millimeters. Go to Cells and click the pop-up arrow. Choose Edit.

4 Create a custom size
In the New Custom Size window, type dimensions of 35x45mm. Click Add. This creates a cell with standard passport photo dimensions. Once you've created your custom passport size, a 35x45 button will appear. Click it to add more passport pictures to your layout. Drag them into position.

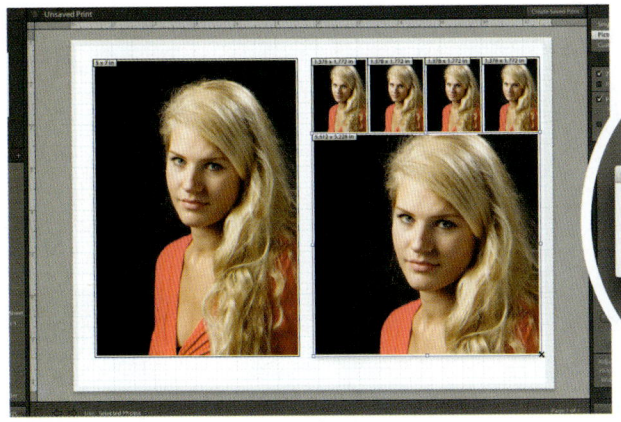

5 Adjust the size manually
Set the Units back to Inches. Click to add a 2.5x3.5 sized cell to the page. Drag the cell's handles to make it fill the layout's empty space. This shape doesn't match the original composition. Hold down Cmd/Ctrl and drag the cell to fine-tune the position of the enclosed photo.

6 Save your template
Clear Zoom to Fill in the Image Settings panel to make the entire photo fit inside each cell. To save the custom layout as a template, click the + icon in the Template browser. Name the template and click Create. You can then summon it with a click in the Template browser.

119

Teach yourself Lightroom

PRINT AND PUBLISH

GET THE FILES HERE: http://bit.ly/tylr2016

Watermark your images

Use the Print module's Page panel to protect your photos by adding watermarks or an Identity Plate to each image

Online galleries and portfolios provide quick and easy ways to share your photographs, but your photos are vulnerable to theft on the internet. By shooting a photo you automatically own its copyright, but once it's been downloaded and re-posted to other sites, prospective clients won't necessarily know who the photograph belongs to. This might result in a photo being used without anyone crediting or paying you. You may also want to send a client proof photos to peruse and order, without the risk of them making unauthorised prints.

As we saw in chapter one, you can digitally assign your copyright and contact details to a photo's metadata, so that this important information will always be attached to the image. This helps potential clients to contact you if they want to use the image in any way. Metadata is useful, but people can still decide to ignore it and make an unauthorised print of your image.

To protect your photos more effectively, Lightroom's Print module has a handy Page panel that enables you to add visual watermarks and Identity Plates to each image. You can create watermarks by adding your own graphic-based logos or by customising existing text. The trick to adding a watermark to an image is to create informative text without obscuring the colours and composition of the image.

1 Add an Identity Plate

We'll follow on from the previous walkthrough and add watermarks to the photos in our custom layout. Toggle open the Page panel on the right. Click the Identity Plate box. Tick Render on every image. Click the preview window and choose Edit to customise the contents.

2 Customise the text

You can click the Use a graphical Identity Plate button to import a logo file. This option supports transparent pixels. Alternatively, tick the styled text option. Type some text such as 'PROOF'. Choose a font from the menu. You can also pick a colour for your custom text. Click OK.

Teach yourself Lightroom
PRINT AND PUBLISH

3 Fine-tune the text
Set the Page panel's Opacity slider to 32% to get a balance between being able to read the text and see the photo. Drag the Scale slider to 62% to enlarge the text so that it can't be removed with a crop. All the photos on the page are now safe to share.

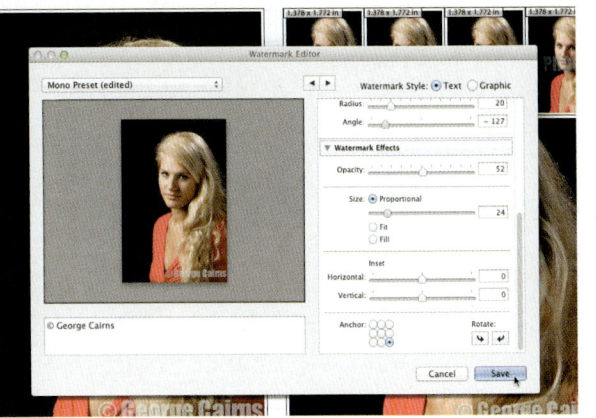

4 Place your watermark
Tick the Watermarking box. You can assign any copyright presets you created following the walkthrough in chapter one. Alternatively, click Edit Watermark and type some text. Choose which corner you want it to appear in. Click Save to create a new preset.

121

Teach yourself Lightroom

PRINT AND PUBLISH

GET THE FILES HERE: http://bit.ly/tylr2016

Publish your photos online

Export your processed pictures to social networking sites such as Facebook, and display photos online with Lightroom Web

One of the most common ways to share pictures is via social media sites such as Facebook and Flickr. These electronic outlets enable us to enjoy instant feedback on our work from friends, colleagues or other people who are interested in photography and image editing. Lightroom acknowledges the need to showcase your images by including upload links to some of the most popular sharing sites in its Publish Services panel. This panel also links to Behance, a service for creative professionals who need to share their images and exchange feedback with their peers. The Publish Services panel also enables you to save exported images in a particular format (such as JPEG or DNG). Simply click on Set Up and log into a service such as Facebook using your password and login details.

Lightroom CC offers another very useful feature — namely, Lightroom Web. This is separate to its Web publishing module, which creates web galleries for uploading to your own website host and requires a little more work. Lightroom Web doesn't require any intervention at all, because any Collection that you synchronise with Lightroom Mobile is also uploaded to your own online space where you can view your collections, share links with others and even edit your images after they're online. To see this in action, simply follow our walkthrough below.

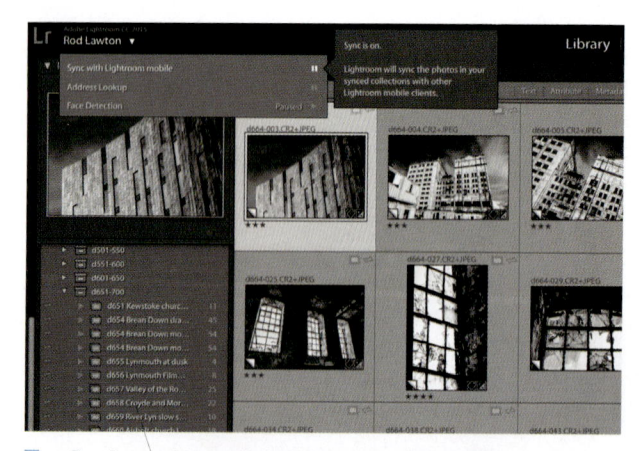

1 Log in to Lightroom Mobile
You may be logged in already, but if not you need to click on the identity plate in the top-left corner of the Lightroom window. You'll be prompted to enter your Adobe username and password, and once you've done that the Lightroom mobile sharing features will become available.

2 Sync collections
To share a collection (you can't share folders or smart collections yet), you can either click the Share box to the left of the collection name, which will display a sync symbol, or use the right-click context menu. You can use the same menu to view shared collections online.

Teach yourself Lightroom

PRINT AND PUBLISH

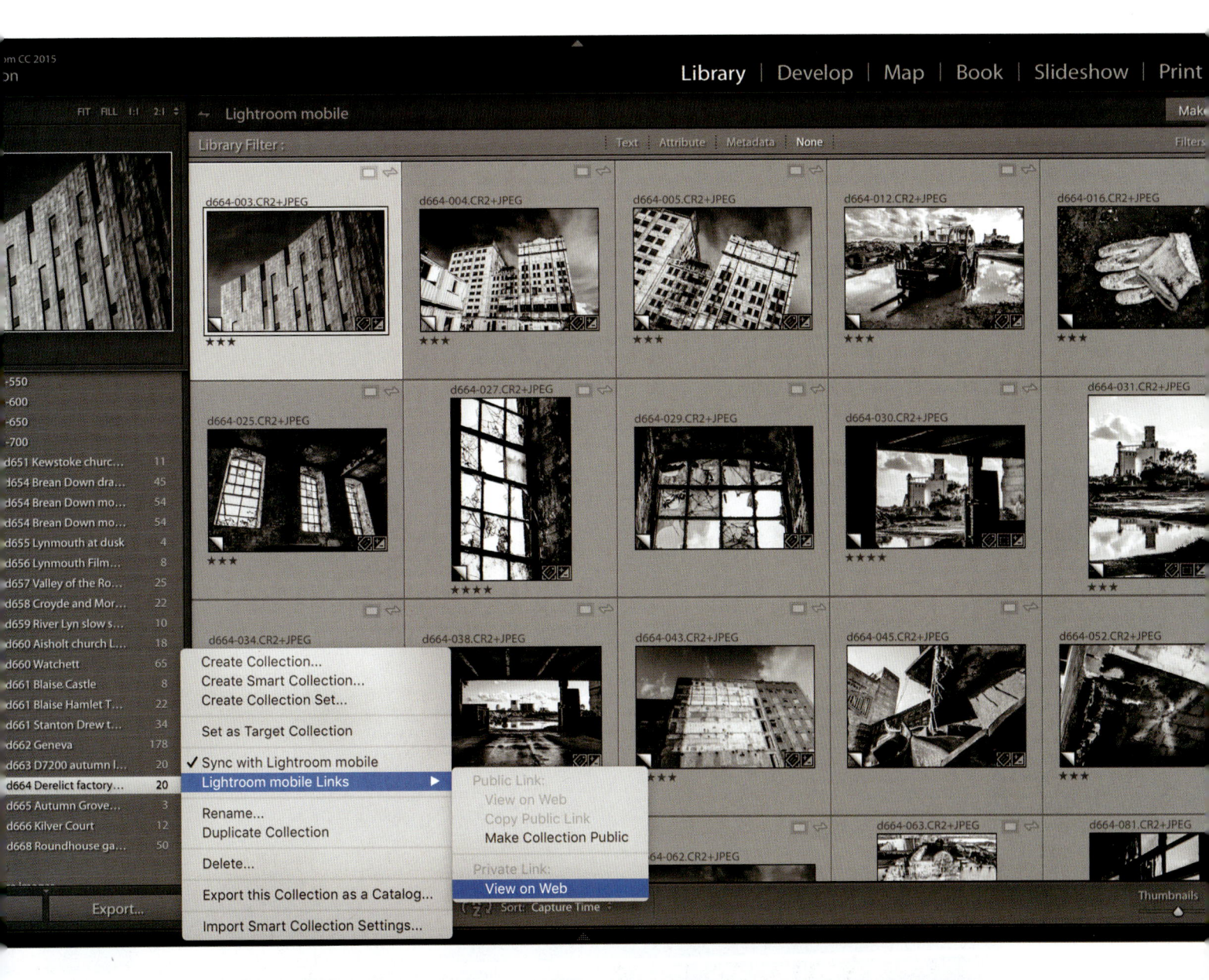

3 View your collections
The View on Web menu option in the previous step image will launch a browser window showing the contents of your shared collection — and all of your other shared collections, displayed as a vertical column down the left side of the screen. Note the Share button at the top.

4 Edit your photos
You can edit your photos within your web browser, using a simplified set of tools in three tabs: Crop, Presets, Adjust. The image we have open here shows the adjustments already applied in Lightroom on our desktop. Any changes we make here will be synchronised with that.

Teach yourself Lightroom
PRINT AND PUBLISH

GET THE FILES HERE: http://bit.ly/tylr2016

Create an online photo portfolio

Showcase your pictures in an interactive gallery using the tools and templates in Lightroom's Web module

Many photographers present their images in an online gallery to promote their shooting and image-editing skills, enjoy feedback, and drum up business. You may find that although you have a website showcasing your work, it's out of date. This could be due to the fact that it can be time-consuming to work on a website's content while trying to master web design packages. You may also have been distracted by easier ways to share and your photos, such as by using a Facebook page, where no HTML or Flash knowledge is required.

Lightroom's Web module enables you to take a Collection of images and turn them into interactive web galleries without any coding knowledge. In this walkthrough we'll show you how to explore a range of web gallery templates and preview how your images will look in a web browser. Lightroom automatically resizes your photos to enable them to be uploaded and displayed online, as well as creates all the files necessary to make your site function. You can use the Web module to add text and email links, so viewers of your site will be able to learn more about your work and get in touch with you.

You can also use the watermarking techniques featured in the previous chapter to make sure that your online images will be protected from misuse, even if they are downloaded from your site. While Lightroom's Web module can help you design your website and create all its web-ready component files, you'll still need the services of a web hosting provider such as www.1and1.co.uk so that you can upload your Lightroom gallery to it.

Teach yourself **Lightroom**
PRINT AND PUBLISH

Teach yourself Lightroom

PRINT AND PUBLISH

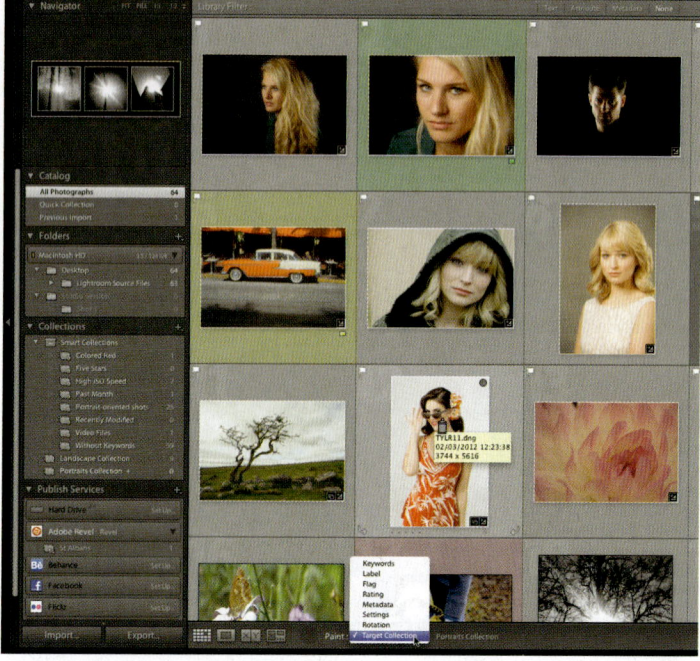

1. Create a Collection
As with creating photo books or slide shows, it makes sense to gather the contents of your web gallery to a Collection before moving them to the Web module. This time we'll use the Painter tool to collect our portrait-themed images together. In the Library module, go to the Collections panel and click the + icon. Choose Create Collection. Label it Portraits Collection. Tick the Set as Target Collection box. Click Create.

2. Add to the Collection
Click the Painter tool icon. Set its drop-down menu to Target Collection. Click any portrait photos in the Grid view. Clicking on a photo will add it to the Target Collection (which in this case is the Portraits Collection). You'll see the number value by the Portrait Collection's label increase as you click new portrait thumbnails to add to it. The Painter can be set to adjust a variety of attributes, such as adding a specific star rating, for example.

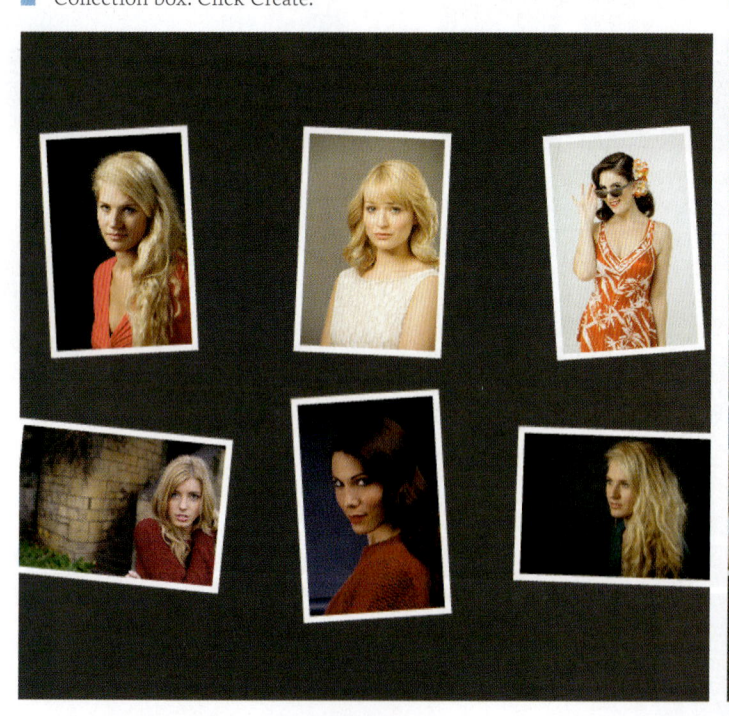

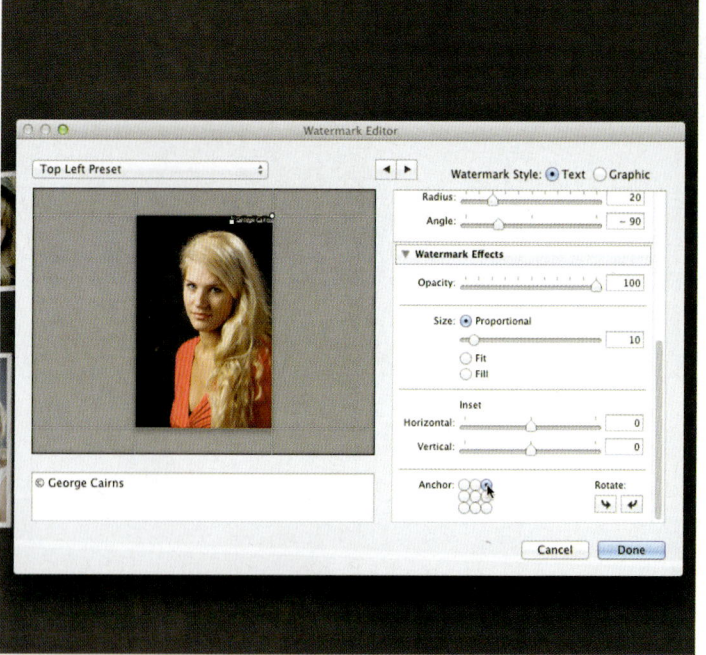

3. Modify the appearance
Click the Portraits Collection label to see your photos. Click Web in the Module picker. Go to the Layout Style panel and choose Airtight Postcard Viewer. The gallery preview window will display your collection as thumbnails. Click a thumbnail and it will zoom in to fill the screen. Use the Appearance panel to choose the number of columns that suit your Collection. Increase the Distant Zoom factor to enlarge the thumbnail size.

4. Add watermarks and captions
As with the Book module, you can add watermarks and captions to your photos. In the Image Info panel tick Caption and choose Exposure to add the camera settings metadata to each photo. In Output Settings tick Watermarking and apply a watermark preset (such as the one we created in the previous chapter). To stop the watermark overlapping the photo's metadata caption at the bottom-left, click Edit Watermarks and change the anchor point to the top-right.

Teach yourself Lightroom

PRINT AND PUBLISH

5 Save the gallery
When you click a thumbnail it will fill the screen and display the camera settings as a caption on the bottom-left. Your copyright watermark will appear at the top-right. If you're happy with your gallery's layout and captions, click the Create Saved Web Gallery button at the top of the workspace. Label your gallery. Click Create. You'll now be able to see and access your saved web gallery in the Collections panel, so you can access and modify any of its attributes.

6 Upload the gallery
To upload the components of your gallery to a hosting site's FTP server, go to the Upload Settings panel and set the FTP Server drop-down menu to Edit. Type in your hosting site's FTP details and enter your user name and password. Click OK. You can then use the Web module's Upload button to get your gallery online. Alternatively, you can click Export to create all the assets needed to present your site and upload them using a third-party FTP application such as Transmit.

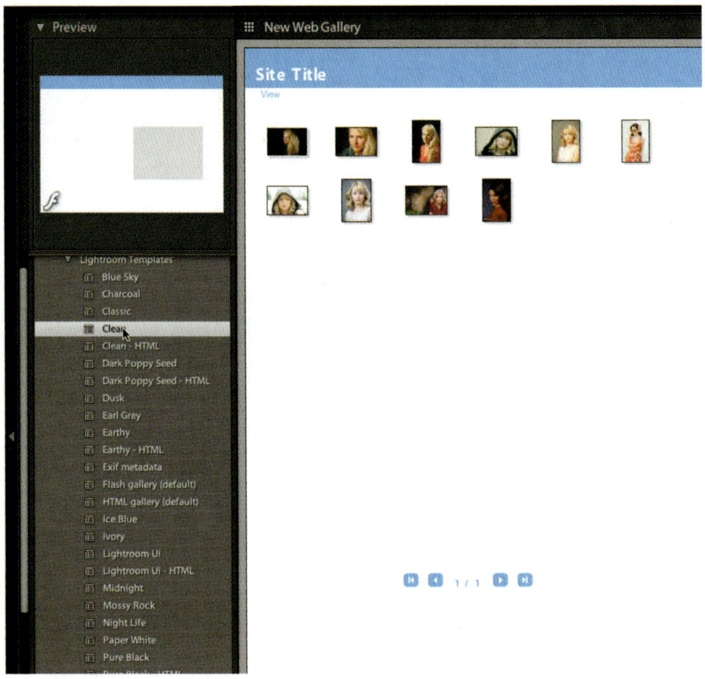

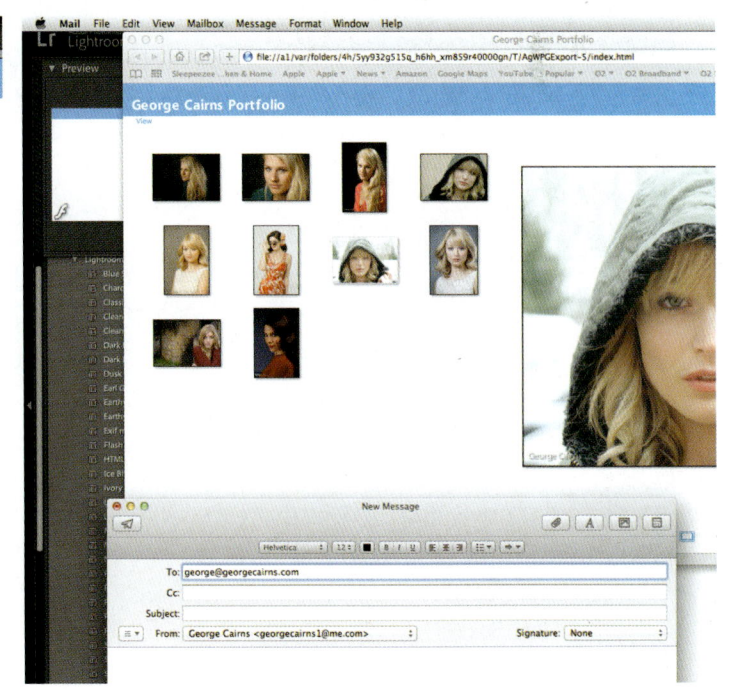

7 Experiment with layouts
Once you've saved a version of your gallery in the Collections panel, experiment with other template layouts. You can access your original gallery at any time. Go to Web>Create New Web Gallery. Label it. Toggle open the Template browser at the left of the Web module. As you move the cursor over the templates you'll see a preview of the layouts in the Preview window. Click to select a layout (such as Clean) to see how your photos look in that template.

8 Add your email information
Type into the new gallery's text fields to add information. In the Site Info box type your email address into the Web or Mail Link field. When people click your name in the gallery this will launch their mail application and automatically fill in your email address. To test how your gallery looks and behaves, click the Preview in Browser button. Lightroom exports the assets and presents them in your browser. You can then check the mail link.

127

Teach yourself **Lightroom**
ADVANCED SKILLS

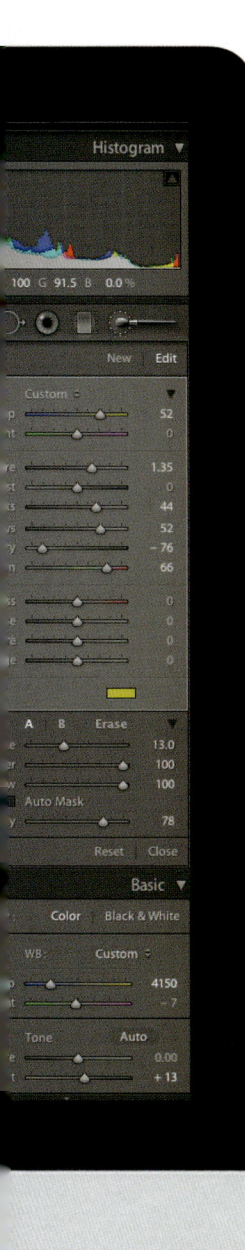

Teach yourself **Lightroom**
ADVANCED SKILLS

Advanced skills

You've learnt the basics, now flex your creative muscles with our extended Lightroom projects for advanced users

130 **Get creative with Lightroom**
Master the Develop module and apply a limitless array of creative effects to your own photos

138 **Retouch your images like a pro**
Polish your portraits using Lightroom's raw-editing tools to apply professional retouching techniques

Teach yourself Lightroom

ADVANCED SKILLS

GET THE FILES HERE: http://bit.ly/tylr2016

Get creative with Lightroom

Master the Develop module for a limitless array of creative effects

Over the next few pages we'll show you a side of Lightroom you may not have seen before. We'll take it for a creative spin and uncover how to use the Develop module (or alternatively, the near-identical controls in Camera Raw) to transform your images in a range of wonderfully creative ways.

Compared with Photoshop Elements or CC, Lightroom doesn't offer anywhere near the depth of tools for image editing. But it does give you the kind of tools you need to make a huge range of tonal effects. What's more, because Lightroom employs parametric editing (where all changes are saved in a 'sidecar' file), everything is completely non-destructive.

Of course, we can use separate layers for non-destructive editing in Photoshop, but the top-down way that layers work means you still have to approach tasks in a certain regimented fashion. By contrast, Lightroom's approach means you're free to change anything, whenever you like. Which in many ways makes it easier to experiment with tonal effects by — for example — converting to monochrome or shifting colours. So for certain tasks, Lightroom offers even greater creative freedom than Photoshop.

Of course, Lightroom (or Camera Raw) isn't a Photoshop replacement. Instead, it should form one part of a photographer's workflow. For most, the best approach is to get as far as possible with the tools in Lightroom (or Camera Raw), then open the image into Photoshop for further changes. But this doesn't mean Lightroom is just for the basic stuff. The tools and sliders within its Develop module enable you to make all kinds of transformative effects, from dreamy colour washes to retro effects, mono presets and crazy make-up. Turn the page to discover how...

Teach yourself Lightroom
ADVANCED SKILLS

Teach yourself **Lightroom**

ADVANCED SKILLS

Creative colour wash effects

Use Lightroom's powerful selective tools to recreate these creative colour effects in your own photographs

You might think that Lightroom is only designed for making basic tonal tweaks, black-and-white conversions, sharpening, and so on. In other words, for all of the fundamental things that photographers need. Lightroom does all of these things really well, but it also gives you the opportunity to transform your images in surprisingly creative ways, such as this colourful portrait effect opposite.

Two powerful Lightroom tools central to this technique are the Graduated Filter and Adjustment brush tools. Both of these tools enable you to define an area for adjusting – either by painting, or by plotting two points for a gradual blend.

You can then use the sliders at the top-right of the interface to change the exposure, colour or detail as you see fit. The most obvious use for this is for selective lightening and darkening, but it's also possible to create interesting effects by toying with colour, saturation and clarity settings. The walkthrough on the next spread takes you through the steps required to recreate the image on the right, but the principles remain the same for whatever image you use, so feel free to experiment with your own images. We also reveal four further Lightroom treatments you can try for yourself...

BEFORE

Teach yourself **Lightroom**

ADVANCED SKILLS

AFTER

Teach yourself Lightroom
ADVANCED SKILLS

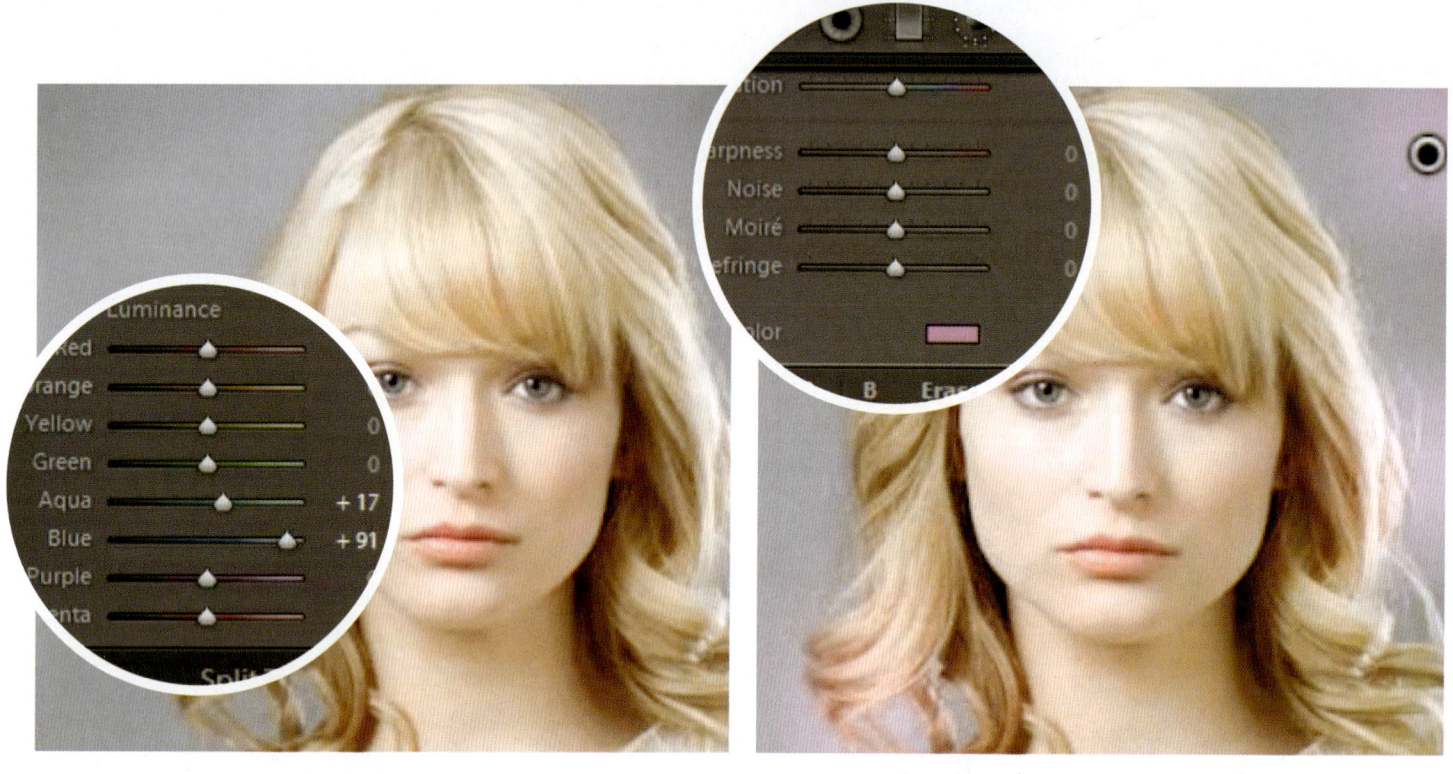

1 Lighten the background
Import 'cover_before.dng' into Lightroom, then go the Develop module. Set Temperature to 4500, Tint to 0, Exposure to +0.36, Highlights to -14, Blacks to -21, and Vibrance to +26. Next, go to the HSL Panel and click Luminance. Grab the Target tool from the top left of the panel, then click and drag upwards on the background to increase the Blue and Aqua sliders. Grab the Adjustment brush from the toolbar.

2 Paint the coloured spots
Scroll up to the Adjustment brush sliders at the top of the right-hand side panels. Click the Color box and choose a pastel pink colour, then increase Exposure slightly to about 0.69. Go to the Brush settings and set Size 17, Feather 100. Click in the image to set an Edit pin, then click around to add spots of subtle colour, using the square bracket keys to resize your brush tip as you go.

3 Add more colours
Click New at the top, then go to the Color box and choose a different colour, keeping to subtle pastel tones. Click around the image a few more times to add different colour spots, building up the effect gradually. Click New again to set more Edit pins for other colours. You might want to try sticking to pastels of one or two colours, such as red and purple or blue and green, rather than going for the pastel neon look we've gone for.

4 Apply coloured gradients
Grab the Graduated Filter from the toolbar. Set similar colour and exposure settings as before, then drag in from the top-left corner to add a gradated blend that starts off in the corner and gradually falls off. Click in other areas to add more coloured gradient effects, adjusting the colours as you go. It's best to build up the layers of effects subtly. Take Snapshots as you go, so you can easily compare different versions.

Teach yourself Lightroom

Four Lightroom treatments

Work fast with a tool that combines selection features with useful preset image adjustments

BEFORE

STANDARD RETOUCH

Use the Basic Panel sliders to increase colour and reveal detail, then plot an S-shaped curve on the Tone Curve line to add extra punch and saturation. Next, grab the Adjustment brush, paint a mask over the eyes and use the sliders at the top-right to lighten and boost the area. Click New and paint another mask over the skin, then set Clarity to -70 to soften the area.

POP COLOUR

Go to the HSL panel and choose Saturation, then simply grab the Target tool from the top-left of the panel and drag down over the colours you want to remove, or use the sliders in the panel. Next, grab the Adjustment brush and set Saturation to -100, then paint with the brush to completely remove any remaining unwanted colour.

RETRO FILM EFFECTS

Crop the image to square then grab the Graduated Filter tool. Set Exposure -4.00 and draw four thin gradients around each edge. Drag another gradient in from one corner, then click the Color box and choose a bright orange, with Temp 72, Exposure 2.25 for a light leak effect. Go to the Tone Curve, click Channel and select Blue. Drag up on the bottom-left corner and down on the top-right.

CLOWN FACE

Go to the HSL panel and click Hue, then drag Blue to +93. Grab the Adjustment brush and paint a series of masks while using the tonal sliders and Color box to alter different parts of the face. For example, to add the mouth, we've painted a mask over the area, then set Temperature -40, Exposure -1.85, Clarity 100, Saturation 90, Color Red.

135

Teach yourself **Lightroom**
ADVANCED SKILLS

Instant mono magic effects
Apply quick, easy monochrome effects with Lightroom presets

Lightroom is a preset lovers' paradise. Presets are so easy to set up, organise and apply that you could conceivably never have to perform the same task more than once — just set the task up as a preset and apply forever after with a single click. We've used a few of the existing Black and White and toning presets here to give you an idea of the effects that can be quickly applied. To try them out for yourself on your own images, simply click through them in the Presets panel to the left of the interface. Even if the Presets don't have the exact desired effect that you were looking for, they can be a useful starting point for further tweaks using the right-hand side panels. And if you do decide to make a few changes, why not set up an alternative preset of your own? Just click the plus icon in the Presets panel when you're done, to save it for future reuse. ■

BEFORE

SPLIT TONE 1

GREEN FILTER

Teach yourself Lightroom
ADVANCED SKILLS

SPLIT TONE 4

RED FILTER

CYANOTYPE

Teach yourself Lightroom

ADVANCED SKILLS

GET THE FILES HERE: http://bit.ly/tylr2016

Retouch your images like a pro

Polish your portraits using Lightroom's raw-editing tools to apply professional retouching techniques

Many of us consider Adobe Photoshop to come in two versions — the budget-friendly Photoshop Elements and more sophisticated Photoshop CC. We tend to overlook Photoshop Lightroom as something that falls between these two stalls. Lightroom CC is now sold in a package with Photoshop CC, equating it in some people's minds with the cheaper Elements, or even with a freebie. But it's actually just as powerful as the more expensive incarnation of Photoshop when it comes to processing raw files. By shooting in raw you get access to more of a photograph's colour and tonal information, so it's much easier to retouch a shot while keeping blocky JPEG compression artefacts at bay.

Lightroom doesn't give you access to layers, of course, so it's of little use if you want to make a creative composite image. However, if you're primarily interested in fixing colour or tonal problems in your raw files, then Lightroom provides you with a powerful digital darkroom thanks to non-destructive and selective editing tools such as the Adjustment brush and the Gradient tool.

Building on our portrait editing piece in chapter eight, in this tutorial you'll learn how to flatter the subjects of your portraits by removing spots and blemishes, and how to selectively soften skin using Auto Masking tools. We'll show you how to tease out delicate iris colour and texture using the selective Adjustment brush, and how to use the Graduated Filter tool to adjust the lighting of your studio backdrops, which will help to draw the eye to your subject.

You'll also discover that Lightroom is packed full of presets that enable you to perform common photo retouches with a few clicks. Here's how…

BEFORE

… Teach yourself **Lightroom**
ADVANCED SKILLS

AFTER

139

Teach yourself Lightroom

ADVANCED SKILLS

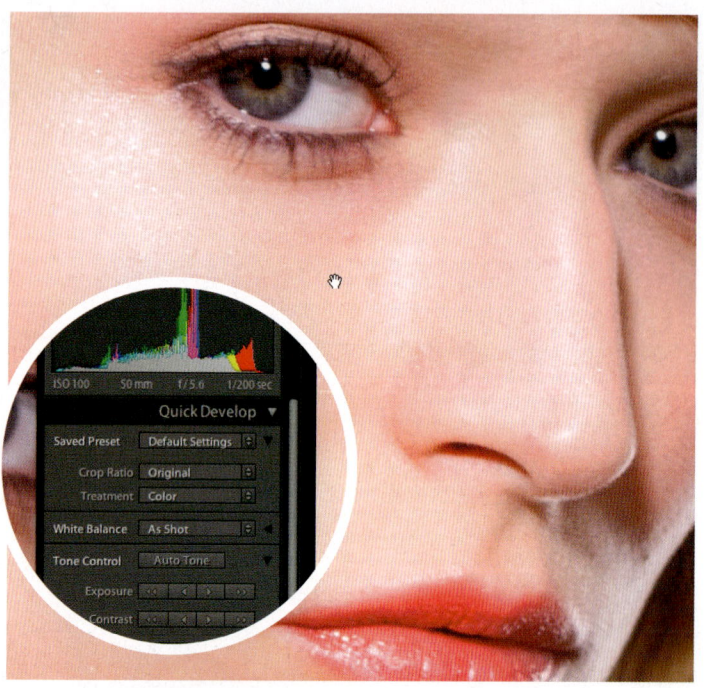

1. Import the starting image
Before firing up Lightroom, download lr_retouch.dng onto your computer's hard drive. When you make edits to this digital negative file you need to be able to save any changes with the file, so it has to be stored on a hard drive (and not a CD) for this to happen. Launch Lightroom. In the Library module, click Import and browse to lr_retouch.dng. Click Add at the top of the window, then click Import.

2. Take a closer look
The unprocessed starting image is a little too warm and has a magenta colour cast. If you tap the space bar to zoom in you'll see a few spots on the subject's face that her make-up has failed to conceal. The focus on her eyes is a little soft too. To retouch the colour, sharpen the eyes and tweak the tones you'll need to enter Lightroom's digital darkroom. To do this, click the Develop module link at the top of the workspace.

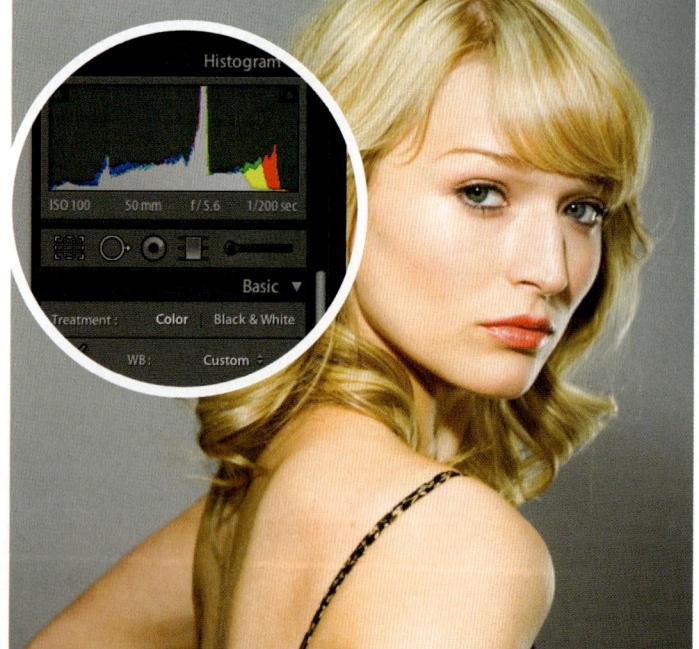

3. Tackle the colour problems
Hit the space bar to see the shot as a whole. Toggle open the Basic panel on the right. In the WB (White Balance) section, drag the Temperature slider left to 5500 to cool things down a little. If you can't get the precise value using the slider alone, you can type it into the field on the right, although it's usually better to use the slider so you can gauge it just right. To counteract the magenta hue, drag the Tint slider left to -5.

4. Tweak the tones
In the histogram window the graph peters out before it reaches the far right. This indicates that the shot lacks strong highlight information. The shadows on the far left are fairly weak too. For stronger shadows and brighter highlights, go to the Tone section of the Basic panel and drag Contrast to 31. The histogram graph spreads a little wider to indicate that more shadow and highlight information is present in the image.

Teach yourself **Lightroom**

ADVANCED SKILLS

5 Boost the midtone contrast
From the histogram you can see that there's lots of midtone information peaking in the middle of the graph. If you drag the Clarity slider right to a value of +31 you can increase the contrast of these midtones without altering the brightest highlights or the darkest shadows. This darkens the midtone grey studio backdrop a little, making the lighter-toned model stand out more effectively within the frame.

6 Adjust the colours
In the Presence section of the Basic panel, drag Vibrance up to +22. This boosts the strength of weaker colours such as the skin tones, without over-saturating the stronger colours in the subject's lips. To fine-tune the overall colour strength, drop Saturation down to -7. These basic tweaks produce more natural and healthy-looking colours and tones in this image. You'll need to experiment with your own pictures.

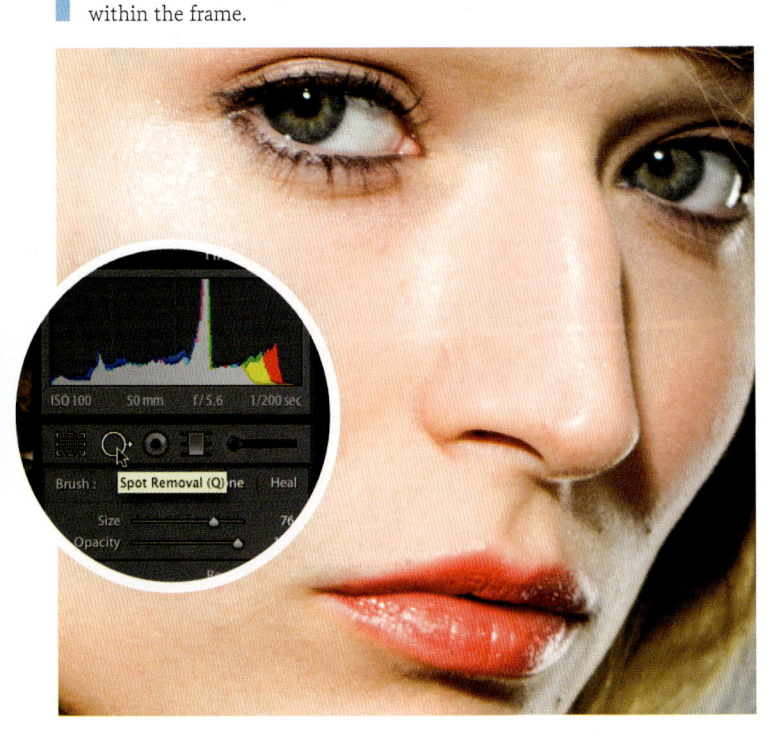

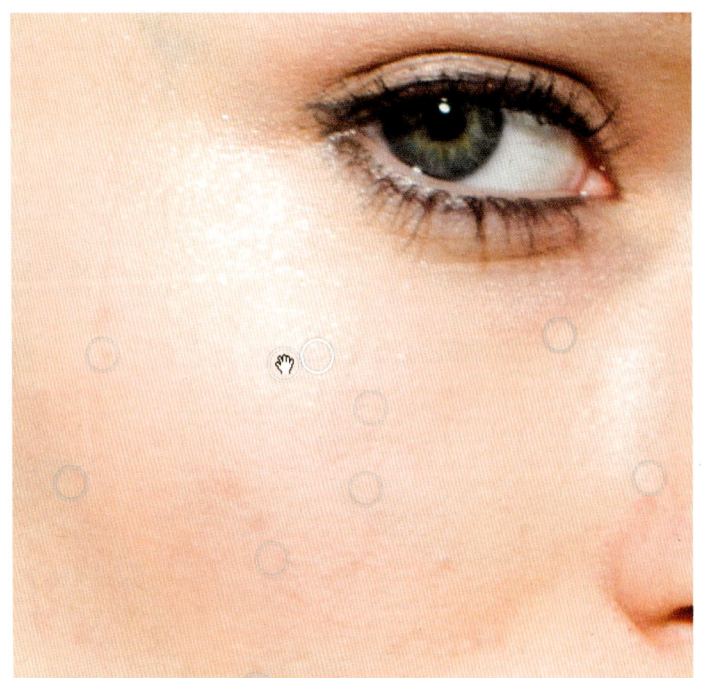

7 Zoom in
Once you've adjusted your portrait's colours and tones using the Basic panel, it's time to focus on enhancing the subject's skin. Click the Spot Removal tool's icon in the mini toolbar (just below the histogram). You can also summon this tool by pressing Q. A new panel will drop down. Click the Heal button, leave Opacity at 100, then press Cmd/Ctrl and the + key to zoom in for a closer look at the tiny details on the face.

8 Remove the spots
Reduce the Spot Removal tool's Size to around 20 by dragging the Size slider left, or by tapping the left square bracket key a few times. If your mouse has a scroll wheel you can use that to resize brush-based tools like this as well. Click a spot or any other blemish on the skin. The tool will automatically sample a clean patch of adjacent skin and place it over the unwanted spot.

141

Teach yourself Lightroom

ADVANCED SKILLS

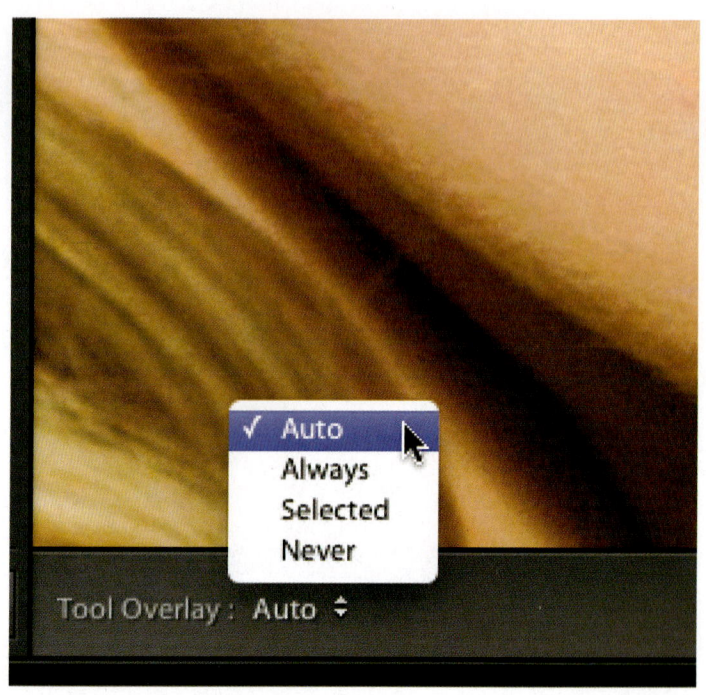

9 Fine-tune your edits
On the whole, the Spot Removal tool will do a good job of automatically sampling a clean area to place over an unwanted spot. However, if the spot is near a detailed feature such as the edge of the face, you run the risk of sampling this area and placing it over the spot. If this happens, you can click the sampled circle and drag it into a more appropriate area to fine-tune the results.

10 Try Tool Overlay
It can be a challenge to see how the spot-removal process is shaping up because of the tool's distracting circular overlays. Navigate to the Tool Overlay option at the bottom-left of the image and set it to Auto. The circular overlays will vanish, enabling you to see the healed areas more effectively. When you move the Spot Removal tool back onto the face, the overlays will reappear, so you can continue to fine-tune them.

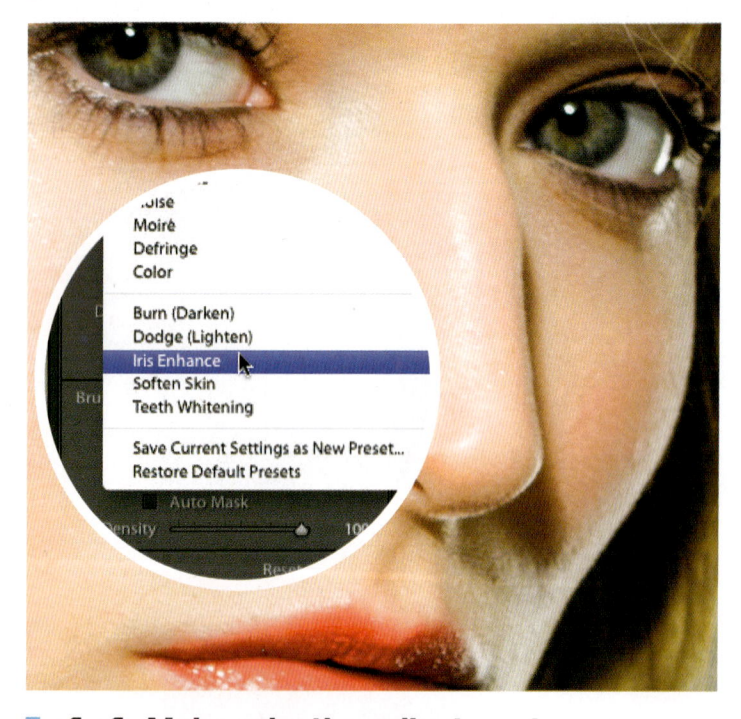

11 Make selective adjustments
Zoom in for a closer look at the subject's irises. They are slightly soft. To sharpen them up without altering the correctly focused areas in the rest of the shot, grab the Adjustment brush from the mini toolbar (or summon it by pressing K). A new panel of editable options will appear. At the top of the panel of sliders is an Effect drop-down menu. Open the menu by clicking Custom, then choose Iris Enhance.

12 Enhance the irises
The Iris Enhance preset boosts the Exposure slider to 0.35 to brighten the eyes. It also increases Clarity to 10 to increase the midtone contrast and reveal delicate iris textures and details. The Saturation slider is increased to 40 to boost the iris colour. Reduce the brush tip Size to 3.0. Click to place an Edit pin on the iris. Now paint to tease out the delicate iris textures and colours. Paint over the other iris too.

Teach yourself **Lightroom**
ADVANCED SKILLS

13 Use masks
If you tick the Show Selected Mask Overlay box at the bottom of the window, the areas altered by the Adjustment brush will appear as red patches. If the brush strokes have strayed over the whites of the eyes, hold down the Alt key and a minus icon will appear in the brush tip. You can now paint to remove certain features from the adjusted areas. Untick the Mask Overlay box when you've finished tidying up the masks.

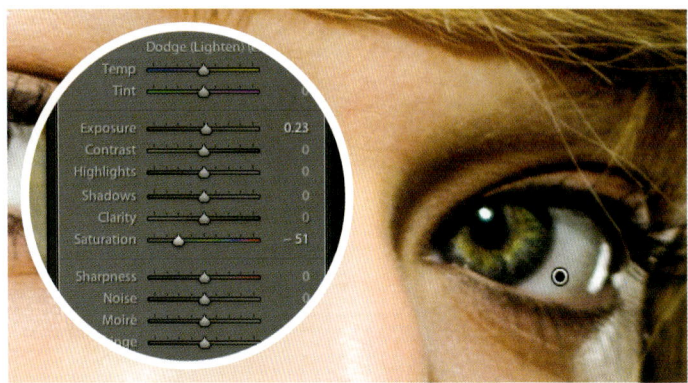

14 Enhance the whites
In the Adjustment brush's properties panel, click New to create a new Edit pin. Now click the Effect drop-down menu and choose Dodge (Lighten). This sets Exposure to 0.25. Click to place an Edit pin on the white of the right eye, then paint over the whites of both eyes to gently brighten them. To make the veins in the eyes look less red, drop Saturation to -51 and paint over these too.

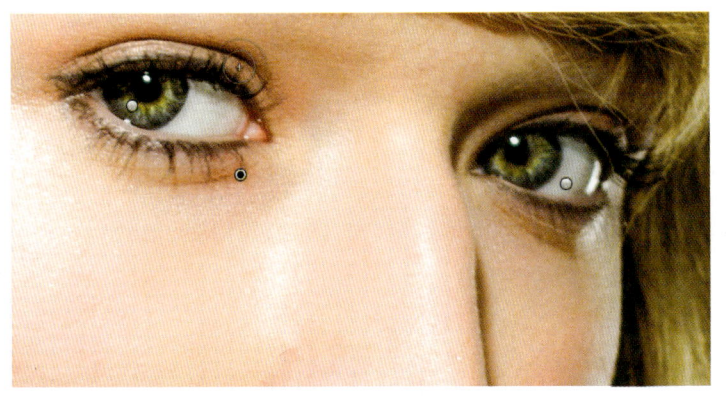

15 Sharpen the eyelashes
Click New to retouch the under-eye areas independently of the eyes themselves. Push the Shadows slider up to 45 so that the brush can selectively lighten the shadows and midtones. Click to place an Edit pin under an eye and then paint to lighten the darker areas and make them less prominent. Click New and set the Effect drop-down menu to Sharpness. Paint carefully over the eyelashes to give them a little more definition.

16 Enhance the lips
Click New once again and choose Saturation from the Effect drop-down menu. By default this pushes the Saturation slider up to 25 so you can boost the colour of specific areas. Click to place an Edit pin on the lips and then paint over them. For a change of lipstick colour, click the Color icon. Choose a colour from the tab or type in a Hue (H) of 21 and set Saturation (S) to a subtle 18%.

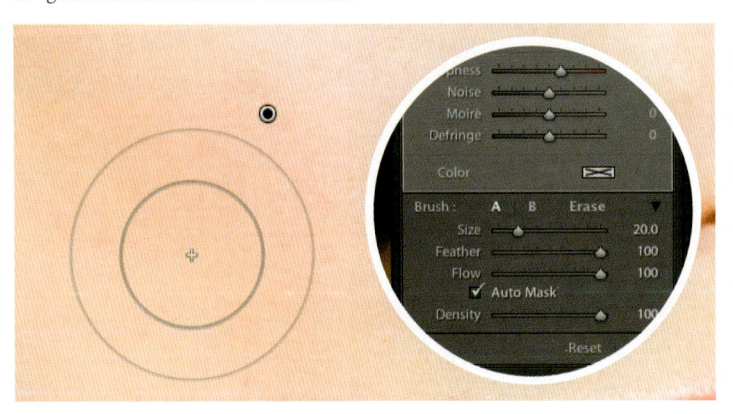

17 Soften the skin
Click New, then choose the Soften Skin preset from the Effect menu. In the Brush tab, tick Auto Mask and set Flow to 100. Click the model's cheek to sample this skin tone and then paint over her face. Similar toned pixels will become softer while other areas will be left untouched thanks to Auto Mask. Click the pin and drag left to reduce the strength of the Clarity and Sharpness sliders for a more subtle skin softening.

18 Apply the finishing touches
Go to the HSL tab and click Luminance. Set Red and Orange to +20 and Yellow to +29. To create a cooler background, go to the Basic tab and reduce Temperature to 4450. Set Vibrance to +44 and Saturation to -24. For more contrast, go to Tone Curve and set the Point Curve to Medium Contrast. Grab the Graduated Filter and set Effect to Burn (Darken). Set Exposure to -1.24. Finally, drag to darken the top-left corner of the image.

143

Teach yourself **Lightroom**
LIGHTROOM MOBILE

Teach yourself **Lightroom**

LIGHTROOM MOBILE

Lightroom Mobile

Discover how you can use Lightroom Mobile on your iPad to sync, edit, and publish your photos

146 Sync Lightroom Mobile
Sync Lightroom Mobile with your desktop copy of Lightroom so you can snap a shot on an iPad and send images to your desktop app's Library

148 Working with Lightroom Mobile
Create a collection of photos in Lightroom and watch it automatically appear in Lightroom Mobile on your iPad and vice versa

150 Sort your Library with Lightroom Mobile
Use Lightroom Mobile to add ratings and flags to your photos, and to sort them using a variety of criteria

152 Shooting super images with Lightroom Mobile
You can take stunning photos on your smartphone using the camera module built into Lightroom Mobile

156 Lightroom Mobile's powerful editing tools
With the latest update, Lightroom Mobile has become a robust editing tool in its own right - even on your smartphone!

Teach yourself Lightroom
LIGHTROOM MOBILE

GET THE FILES HERE: http://bit.ly/tylr2016

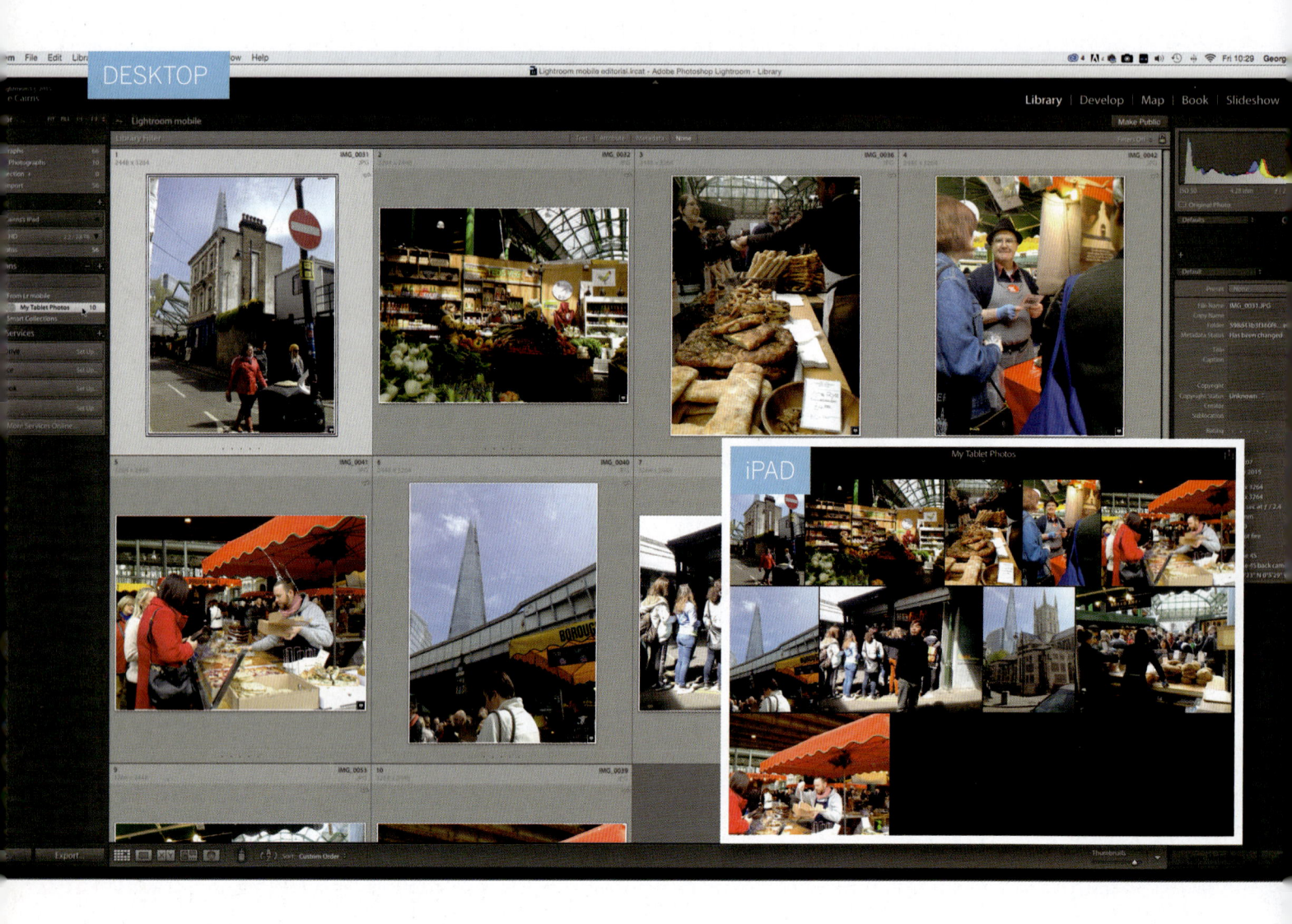

Synchronise Lightroom Mobile

Sync Lightroom Mobile with your desktop copy of Lightroom so you can send mobile images to your desktop app's Library

Lightroom Mobile is an app that will run on a tablet or smartphone. It's a companion app to the full version of Lightroom CC that runs on your Mac or PC. The mobile and desktop versions of Lightroom work seamlessly to enable you to organise, edit and share your photos from your home or on the go.

To use Lightroom Mobile you'll need to sign up for an Adobe ID so that you can access the Adobe Creative Cloud. This enables you to create a collection of photos on your desktop copy of Lightroom and sync them via the Creative Cloud to Lightroom Mobile on your portable devices.

You can then use Lightroom Mobile while on the go to add ratings to your photos and adjust their colour, tone and composition. Once you hook up your iPad to a Wi-Fi connection, the changes you've made to your photographs in Lightroom Mobile will be synced via the Creative Cloud to the photos in your Lightroom Library.

You can also snap shots using your tablet or smartphone's camera and get them to sync automatically to a collection in your desktop copy of Lightroom. Before we show you Lightroom Mobile's sorting, editing and sharing tricks, here's a guide to get both versions of Lightroom to talk to each other.

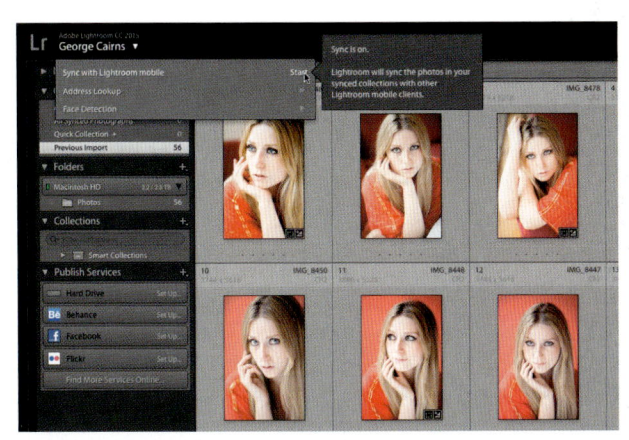

Teach yourself Lightroom

LIGHTROOM MOBILE

1. Turn on synchronisation
In Lightroom CC on your desktop, click the identity plate at the top-left to summon a grey menu box. Click the Start label next to the Sync with Lightroom Mobile command. A new message box will appear saying that Sync is on. Launch Lightroom Mobile on your tablet.

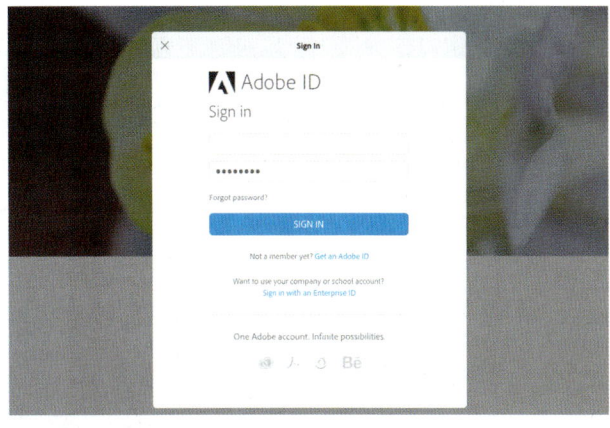

2. Sign in on the tablet
Click the Sign In button on the welcome screen. Sign in to the Creative Cloud using your Adobe ID. When a message indicates that 'Lightroom would like to access your photos', click OK so you can import shots from your device's Camera Roll.

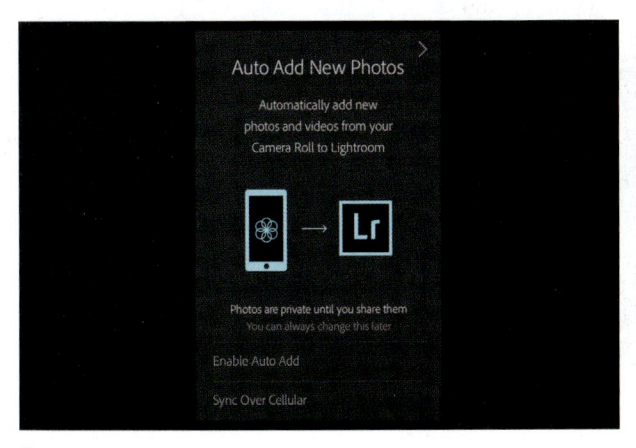

3. Sign in on a smartphone
On the smartphone version of Lightroom Mobile you'll be given the option to choose Auto Add New Photos. Any shots that you snap will then automatically appear in a Lightroom Mobile collection. Untick Sync Over Cellular if you want to avoid exceeding your data plan.

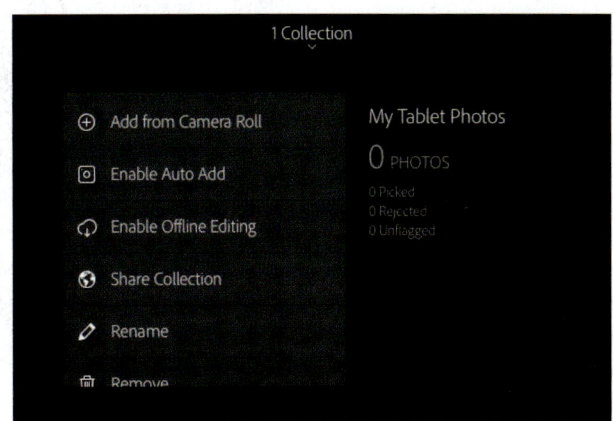

4. Import automatically
In Lightroom Mobile you'll see a blank collection called My Tablet Photos. Tap the three white dots at the bottom-right of the collection's thumbnail to summon a menu with options such as Enable Auto Add. This will import any new photos you shoot on the tablet automatically.

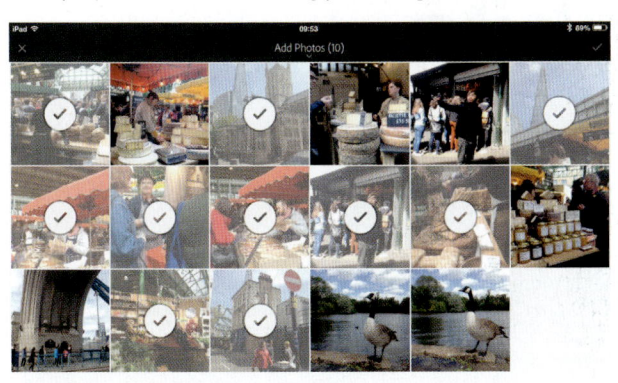

5. Add photos
In the Add Photos window, tap the thumbnails of any iPad (or smartphone) photos you want to import into your Lightroom Mobile collection. A tick will appear on each selected thumbnail. When you've finished, tap the tick at the top-right and the photos will appear in the Tablet collection.

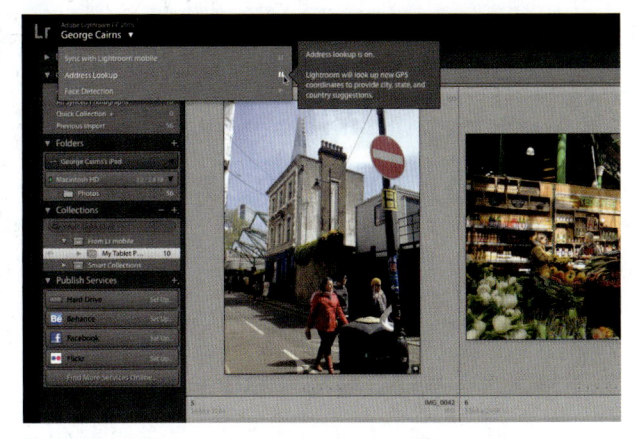

6. View your synced collection
When your mobile device is in range of a Wi-Fi connection it will sync the Tablet collection via the Creative Cloud to your Lightroom Library. In the Enable Address Lookup window, Tick Enable so that Lightroom can pinpoint mobile photos in its Map module.

Teach yourself Lightroom
LIGHTROOM MOBILE

GET THE FILES HERE: http://bit.ly/tylr2016

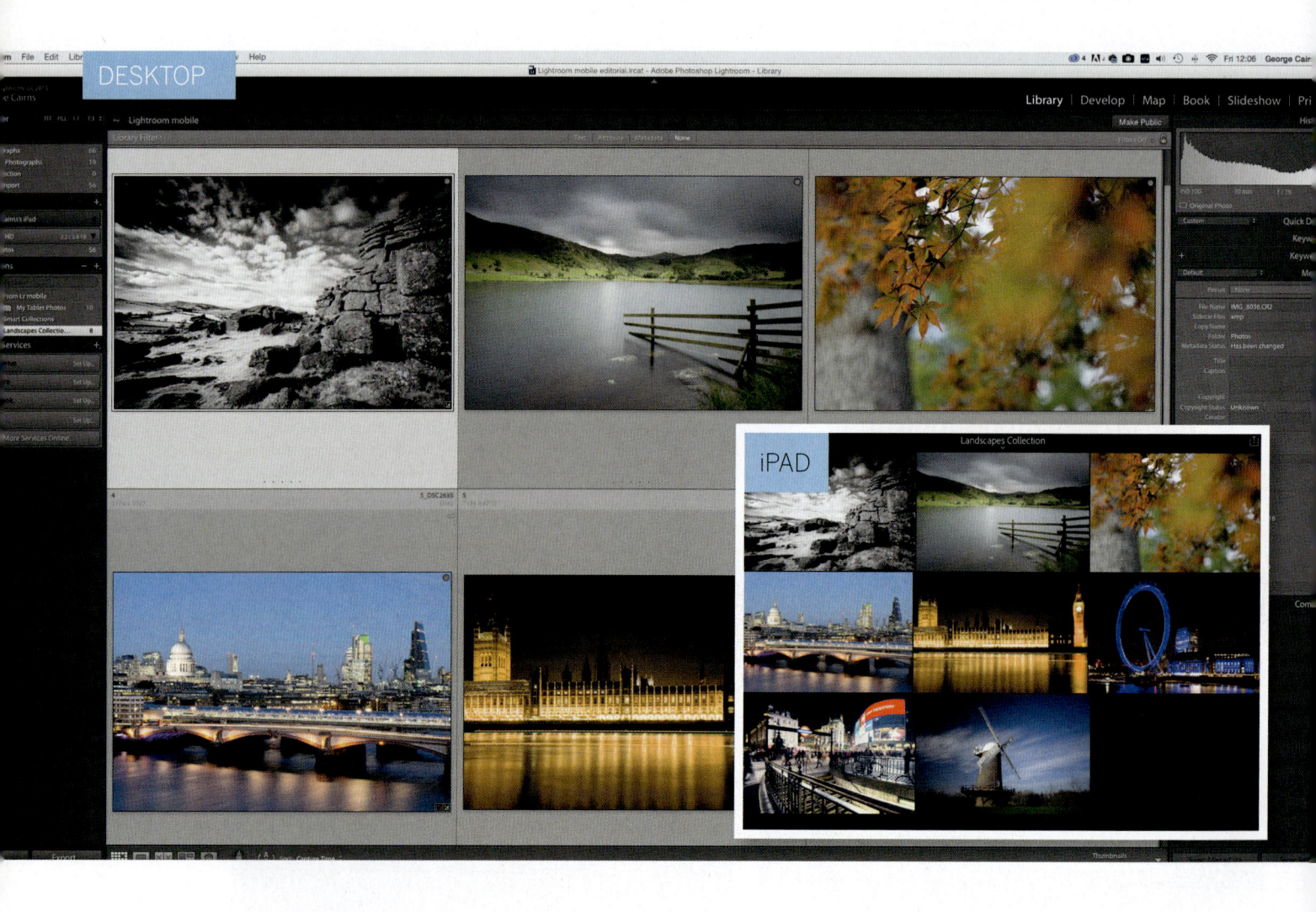

Working with Lightroom Mobile

Create a collection on your desktop version of Lightroom and sync it to Lightroom Mobile on your portable devices

On the previous spread we created our first collection in Lightroom Mobile by importing photographs captured with a smartphone or tablet. By default the collection is called My Tablet Photos, but you can re-label the collection in Lightroom Mobile by tapping the three white dots at the bottom-left of the collection's cover photo. This reveals a list of commands including Rename. You can also rename the collection after it has been synced to Lightroom on your desktop. Go to the Collections panel in the Lightroom Library module and right-click the My Tablet Photos collection label. Choose Rename from the context-sensitive pop-up menu that appears. If you rename the collection in one version of Lightroom, the name change will automatically be updated in the other. As well as syncing photos snapped on your tablet or smartphone, you can also sync photos from your computer Lightroom Library with Lightroom Mobile, so that you can organise and edit them when you're away from your desktop. To do this you need to create collections in Lightroom that are instructed to sync with Lightroom Mobile. Collections enable you to choose precisely which pictures are accessible on both desktop and mobile devices at the same time. Lightroom sends lightweight Smart Previews to your mobile device. When you process these Smart Previews in Lightroom Mobile, the changes are applied to your original Lightroom raw files, enabling you to edit raw files on the tablet.

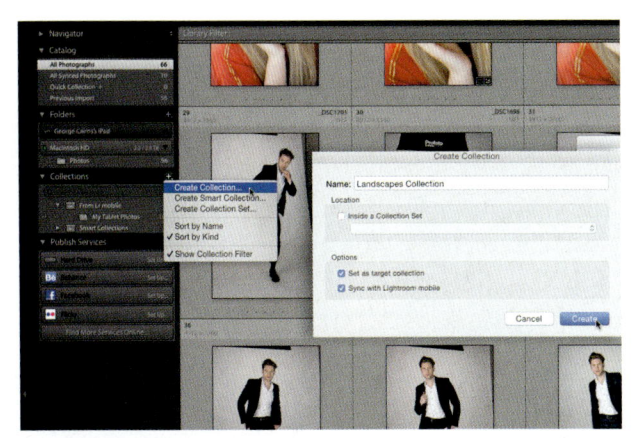

Teach yourself **Lightroom**

LIGHTROOM MOBILE

1. Create a collection
In Lightroom, go to the Collections pane and click the + shaped New Collections icon. Choose Create Collection from the pop-up menu. Label the collection. Go to Options. Tick the Set as target collection box. Leave the Sync with Lightroom Mobile box ticked. Click Create.

2. Add to target collection
To start with, your new collection contains no photos. You can drag and drop appropriate photos from the Library module's Grid view to add them to the collection manually. Or tap a photo's thumbnail to select it and press B. This activates the Add to Target Collection command.

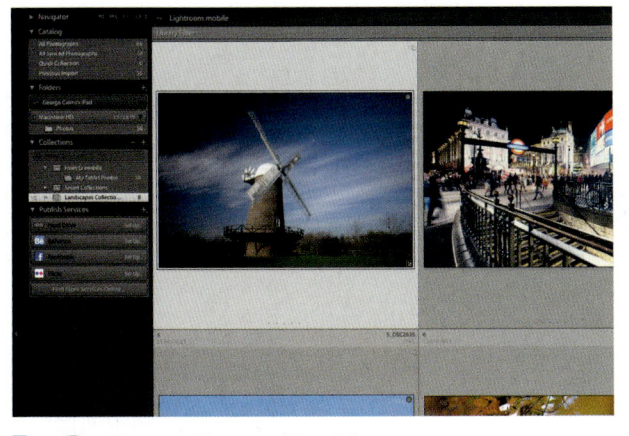

3. Sync the collection
Click the new collection's label in the Collection panel to see its contents. As you add new photos to this collection, a 'Syncing photos' message appears at the top-left of the Library module. This indicates the number of Smart Previews being uploaded to the Creative Cloud.

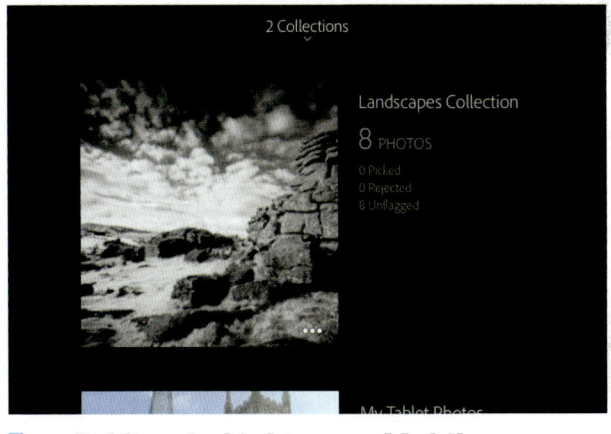

4. View in Lightroom Mobile
Launch Lightroom Mobile on your smartphone or iPad. As well as the original My Tablet Photos collection, you'll see a thumbnail for your new Lightroom-sourced collection (as long as your device is synced to Wi-Fi so that it can access the Smart Previews).

5. Resize the thumbnails
Tap the new collection's thumbnail to view its synced contents. They'll appear in the same order as they appear in Lightroom. You can resize the thumbnails in the collection's Grid view by pinching to enlarge or shrink them. Tap a thumbnail to view the image full screen.

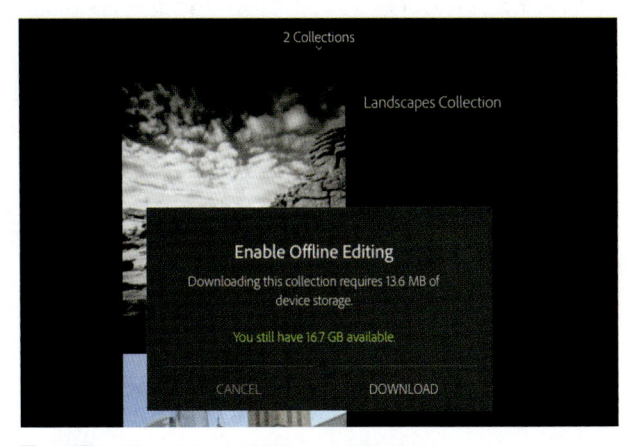

6. Enable offline editing
Lightroom Mobile uses Wi-Fi to show Smart Previews stored in the Creative Cloud. If you want to work on a collection of images without a Wi-Fi connection, go to the collection's thumbnail and tap the three white dots to access a menu list. Then tap Enable Offline Editing.

Teach yourself **Lightroom**

LIGHTROOM MOBILE

GET THE FILES HERE: http://bit.ly/tylr2016

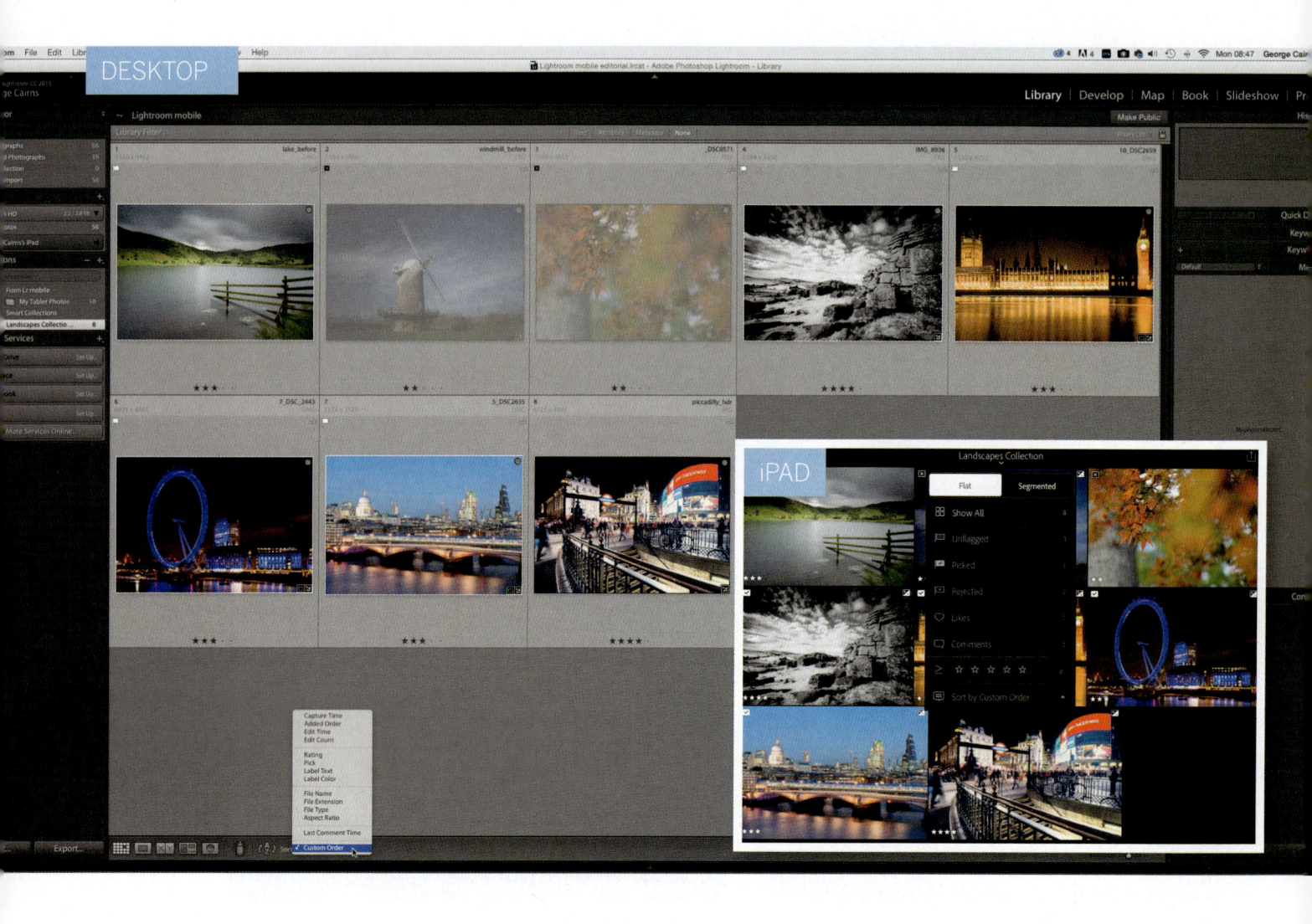

Sort images with Lightroom Mobile

Organise your synced collections by adding ratings and flagging photos to pick or reject them

Once you've created collections in Lightroom or Lightroom Mobile, you can begin to sort your shots to flag up favourites, or mark some files for rejection. You can also highlight images that you're proud of by using star ratings.

If you flag a file or add a rating in Lightroom, then these changes are added to the file's metadata and automatically synced to the version of the photo in Lightroom Mobile's equivalent collection. In the same way, if you rate a photo in Lightroom Mobile, the assigned rating or flag badge will appear by the same photo in the Lightroom collection. This versatility enables you to start rating on your desktop computer and then continue reviewing, rating and flagging your files on your iPad or smartphone. If your tablet lacks access to a Wi-Fi connection, you can still sort your files. The changes that you make will be synced with Lightroom once you connect to Wi-Fi later on.

By adding ratings and flags to a collection of photos you can find particular pictures more easily in the future. Both Lightroom and Lightroom Mobile provide filters that enable you to sort the picks from the rejects, or display files according to specific star ratings. To see any star ratings and flag badges assigned to each thumbnail, perform a two-fingered tap in the Grid view.

Teach yourself **Lightroom**

LIGHTROOM MOBILE

1. Pinch to zoom
In Lightroom Mobile, tap to view the contents of a collection. They are displayed in the Grid view as thumbnails. If you two-finger pinch the Grid view to zoom in, the thumbnails enlarge, making it easier for you to see each photo. You can also pinch to shrink the thumbnails.

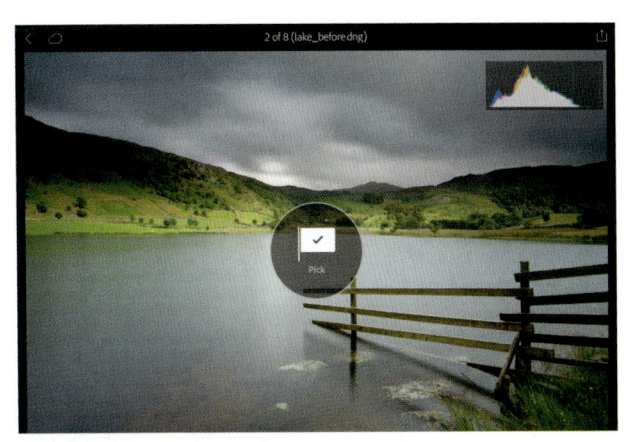

2. Flag your files
Tap a photo's thumbnail to view it full screen. If you click the image and swipe vertically you can summon a flag overlay that features three icons: Pick, Unflag and Reject. Alternatively, tap the three vertical dots at the bottom-left of the workspace and tap one of the three flag icons.

3. Rate your files
Adjacent to the three flag icons you'll see a group of stars. By tapping these icons you can assign a five star rating to favourite files and a lower star rating for less successful images or those that need processing to fix problems with colour, composition or tone.

4. Filter by flag
Once you've flagged or star-rated your photos, tap the arrow-shaped back icon at the top-left of the screen to go back to the collection's Grid view. Click the drop-down icon below the collection's name to open a menu. Tap Picked to only display images flagged with that designation.

5. Filter by rating
Tap Show All. You can also choose to display thumbnails that have a rating that is greater than or equal to a specified number of stars. If you tap the ⌘ icon you can change it to a more specific = icon so that only images with a chosen rating are displayed.

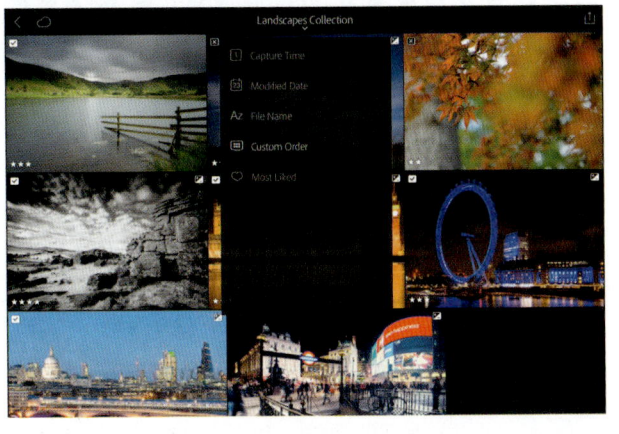

6. Choose a custom order
In Lightroom you can drag a thumbnail image into any position, regardless of its rating. To see this layout in Lightroom Mobile, tap the Sort by Capture Time icon and change it to Custom Order. Lightroom Mobile's thumbs will change position to match the order in Lightroom.

Teach yourself Lightroom

LIGHTROOM MOBILE

GET THE FILES HERE: http://bit.ly/tylr2016

Shooting with Lightroom Mobile

You can take stunning photos on your smartphone using the camera module built into Lightroom Mobile

Although Lightroom users will usually use pretty serious camera kit, many of us also take casual pictures on our smartphones too. In fact, there's a growing art movement in mobile photography.

The camera module built into Lightroom Mobile is really useful. The first reason is that it offers more advanced camera controls that your smartphone's built-in app won't necessarily provide, including white balance and exposure compensation adjustment.

Second is that you can use a small selection of 'shoot-through' presets that apply a specific image effect to your photos, but in a non-destructive way. It's like using regular Lightroom presets in that you can modify the settings or choose a different preset at any time.

The third reason is that any pictures you take with the Lightroom Mobile camera are automatically added to your Lightroom library and synchronised with your Lightroom Catalog on your desktop computer. Brilliant!

Teach yourself **Lightroom**

LIGHTROOM MOBILE

1. Interface
After tapping on the Lightroom Mobile app on your smartphone or tablet and choosing the camera icon in the bottom-right of the screen, the app is ready to go. When first loaded up you're confronted with a live view image of what's in front of your mobile device's lens, along with several buttons on the screen. To the top left is the option to turn the camera around to instead use the device's front-facing camera (for those who want to take a selfie) and at the bottom-left is the option to quit the camera app. Over on the right-hand side we have the self-timer, grid and crop information overlays, exposure, white balance and flash controls. As well as these controls we can also choose to use a preset filter through the overlapping circles icon at the top-right, the shutter button in the middle and image review button on the bottom-right.

2. Information overlays
Click on the grid icon to access the different information overlays. The overlays available in the app contain spirit level, 1 x 1 square crop and grid display options that lay over the live view feed coming from the camera. The spirit level uses your smartphone or tablet's accelerometer to monitor your image's horizontal and vertical alignment, which it displays through a broken centre line and graticule in the centre of the display. This enables you to make sure your shots are level, facilitating straight horizons. The 1 x 1 crop makes it easy to visualise your shot in a square crop, for those of us who want to share our photographs straight to Instagram and other apps and websites. And the grid display lends itself to the stickler photographer inside of us who wants to hit the rule of thirds each and every time.

153

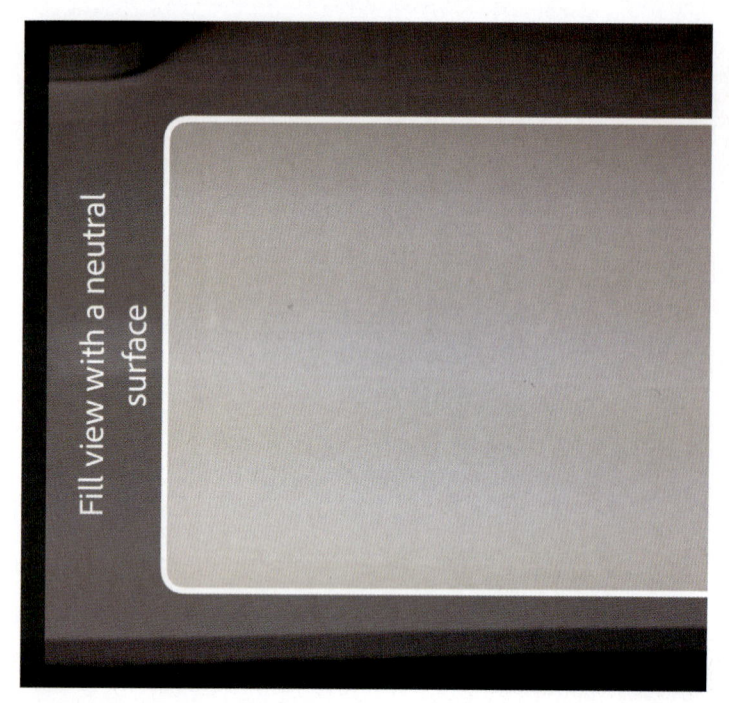

Teach yourself Lightroom

LIGHTROOM MOBILE

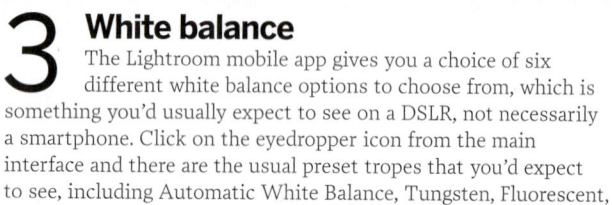

3. White balance

The Lightroom mobile app gives you a choice of six different white balance options to choose from, which is something you'd usually expect to see on a DSLR, not necessarily a smartphone. Click on the eyedropper icon from the main interface and there are the usual preset tropes that you'd expect to see, including Automatic White Balance, Tungsten, Fluorescent, Daylight and Cloudy settings. But the app also has a clever way of getting around custom white balance setup, and that's through the handy Colour Picker option rather than complicated manual adjustments. Simply tap the colour picker option and fill the frame with the colour that you want to set your white balance against (light grey or white smooth surfaces usually work well) and the app does the rest, to enable you to achieve the tone you want.

4. Presets

Found by pressing the icon with overlapping circles that looks like a lens filter being attached to a DSLR, we have five presets to choose from in the app. These are: High Contrast, Flat, Warm Shadows, High Contrast B&W, and Flat B&W. You can immediately see the effect that they give when you tap on each one, because the app applies the filter over the live view feed, enabling you to choose the one that best suits your shot. Essentially these presets alter contrast and colour to manipulate the image into different styles. Try High Contrast to add vibrancy to your photographs, or Flat to create a muted pallette. Meanwhile Warm Shadows increases orange tones to create a warm, sunny look. Blacks look deeper and whites brighter in High Contrast B&W than they do in Flat B&W, so you can create a more modern or a subdued feel to black and white photos.

Teach yourself **Lightroom**

LIGHTROOM MOBILE

5. Flash
Of course Lightroom, like most smartphone camera apps, has a flash function. Click on the familiar lightning bolt flash icon and the options that appear are On, Off or Auto. Set it to On and the flash will always fire each time the shutter button is pressed, regardless of the natural lighting available. The reverse is true for Off. Auto enables the app to decide whether it's dark enough for the flash to fire, and isn't always a particularly reliable judgement because it depends on the meter reading that the app is taking from the subject you're photographing. Low-key photographs suffer from Auto triggering the flash when it isn't wanted, so there are times when you need to override the Auto function by turning the flash on or off. This is easy to do and only takes a second, with the image preview helping you to quickly decide if it's necessary.

6. Exposure
By clicking on the black and white plus and minus icon, you can alter the exposure settings manually through the app by +3 or -3 stops, by scrolling up or down the sliding scale. Again this is previewed in the live view to help you choose. This is useful in most circumstances, but will still struggle with shooting directly into the sun or into a deep, dark cave.

7. Self-timer
The self-timer, selected via the stop watch icon, gives us the option of choosing between Off, 2, 5 and 10 seconds delay. This is perfect if you want to grab a quick snap on your own or with a friend, further than an arm's length from the lens. Pop the camera on the floor or against a wall and press the shutter button. The shutter button then counts down until it takes.

Teach yourself Lightroom
LIGHTROOM MOBILE

GET THE FILES HERE: http://bit.ly/tylr2016

Mobile's powerful editing tools

Lightroom mobile is a robust editing tool in its own right – even on your smartphone!

When you synchronise your images to your mobile devices, you're sharing smaller-sized Adobe DNG-based smart previews that can nevertheless store all of the adjustments that you make. These work in the same was as they do in the desktop Lightroom application – all the changes you make are 'non-destructive' and can be modified or removed at any time.

And this still applies if you start editing an image on your desktop computer version of Lightroom and then carry on modifying the picture on your mobile device in the app. When you look at the image on your desktop computer again, all of the changes that you made on your mobile device will have been added to that version automatically.

Early versions of Lightroom Mobile had relatively modest adjustment features available, but they've steadily evolved into a surprisingly powerful and useful set of image-editing controls, as our walkthrough examples demonstrate. Get to know these features to boost your Mobile editing confidence.

Teach yourself Lightroom

LIGHTROOM MOBILE

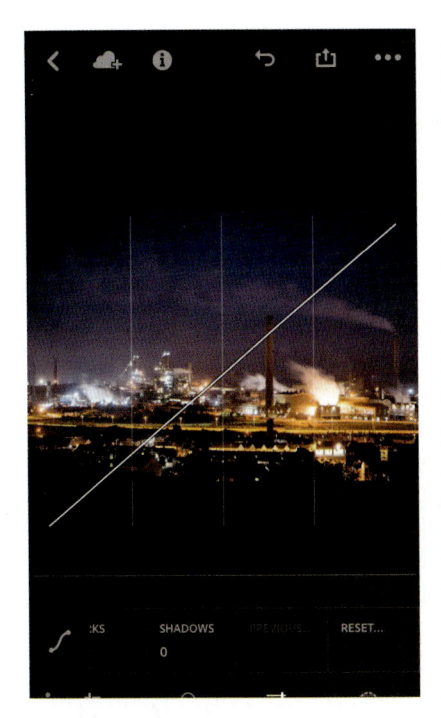

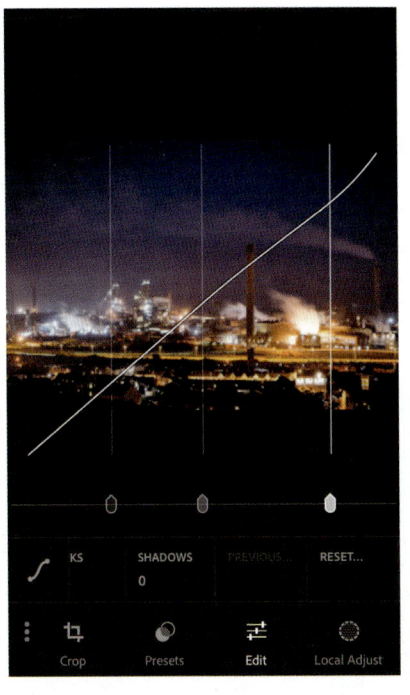

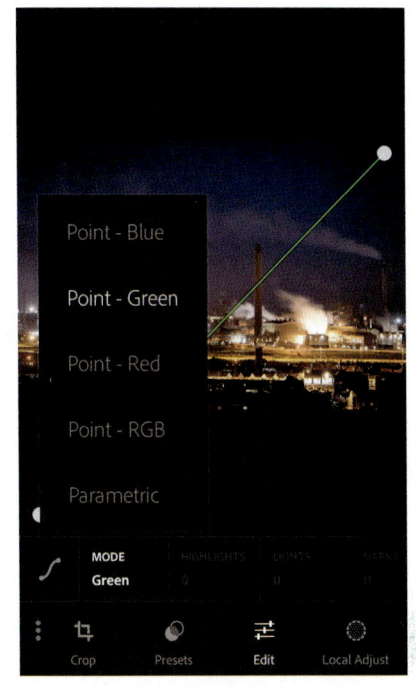

1 Tone curve
The tone curve opens in parametric mode where it breaks down your image into four distinct sections: Highlights, Lights, Darks and Shadows, (in order descending in brightness). If you tap and drag your finger in one of these sections and then slide up and down you'll notice a change in your image. If you want to increase the bandwidth of a section, click the grey sliders below the image and drag them to the left or right. You can be more precise by tapping on the section name along the settings panel at the bottom of the screen and then scrolling left and right on the slider. By tapping on the Mode button you can move from parametric editing to single colour channels: red, green and blue, as well as RGB combined. This is useful for photos with a colour cast.

2 Local adjustments
An excellent addition to Lightroom Mobile is the ability to introduce ND grads. You can adjust temperature, tint and exposure, contrast, highlights, shadows, whites, blacks, clarity, dehaze, saturation, sharpness, noise, moire, defringe, colour hue and saturation for the colour hue. You can choose ND grad selection (named Linear selection) or Radial selection. The Radial selection tool enables you to draw and morph circles and ellipses on your image to affect only that area. Tapping on the bigger control handle and sliding up and down increases and reduces the feathering of the selection. Look to the top of the window and you'll find a plus icon to add another radial selection. If you need to invert that selection, hit the invert button (the circle in a rectangle).

Teach yourself Lightroom

LIGHTROOM MOBILE

3 Split toning

Split toning adds colour control in the shadows and highlights of an image. It enables you to achieve the balance of warm and cool tones that you want, by adjusting the shadows and highlights in isolation from one another. Let's say we want warm shadows to accentuate the streetlights in our image here, and cooler highlights in the sky. To do this, click on Highlights Hue and drag the slider to a blue hue. It looks good at the default Saturation setting of 50, so we'll leave it at that. Next, click Shadows Hue and pick a warm colour, such as orange. This creates quite a dramatic effect, so we'll decrease the Saturation to around 15. If you want to alter the balance between the Highlights and Shadows, tap Balance and move the slider to the left and right to adjust it.

4 Dehaze

A useful new feature that's new to Lightroom as a whole, and available in the mobile app, is the Dehaze tool. It adds even mid-tone contrast and boosts the saturation of images subtly. Dehaze works well on photos with large portions of mid-tone grey, as you would expect to see in a hazy or foggy shot. We can reduce the haze in this rally photograph, for example, by increasing the Dehaze amount to +50. This makes it look brighter and livelier without becoming over-saturated. Or on the other hand, if we want to accentuate the haze effect then we can reduce it to -12 for a more washed-out look, although it doesn't seem to work quite as well at doing this. To add a convincing hazy effect to images, then, you'll need to make further adjustments using other tools.

158

Teach yourself Lightroom

LIGHTROOM MOBILE

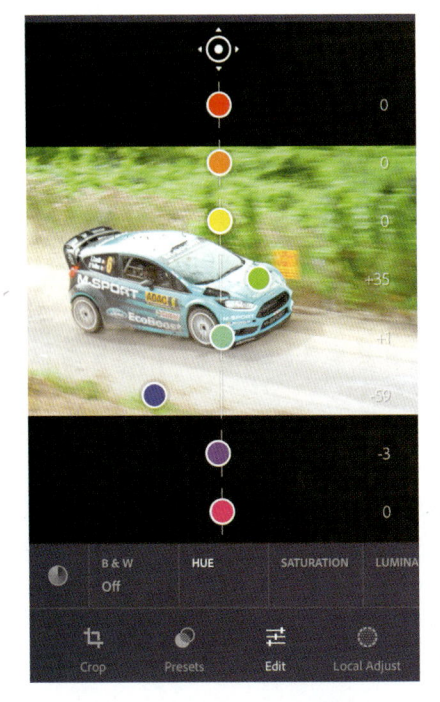

5 Colour and B&W
In the Colour and B&W tab are some intuitive controls to modify colours. Tapping on Hue overlays colourful sliders on your image. Tap and drag left and right to change the hue of a specific colour.

We turned the blues more cyan and the greens more green. Next we reduced the green channel's luminance to darken the foliage behind the car to push focus onto the brighter car. You can adjust saturation on all eight colour channels as well, which can even out skin tones. Tapping on the

B&W button switches the photograph into a black and white photo that we can alter single colour channels of. By boosting a particular colour it makes it brighter (positive luminance) and by attenuating a colour we darken in, meaning we can be more selective than with a brush.

6 Vignette
The vignetting tool boosts those photos that need extra emphasis on the subject. You can apply a black or white vignette with varying intensity by tapping on Amount and moving the slider left for black or right

for white. For a subtle effect we used -20 for a slight darkening around the frame. Midpoint selects where the vignette stops, Feather changes the blur to the vignette's edges, and Roundness morphs its shape. Tapping on the Highlights button displays three functions, Highlight Priority, Color

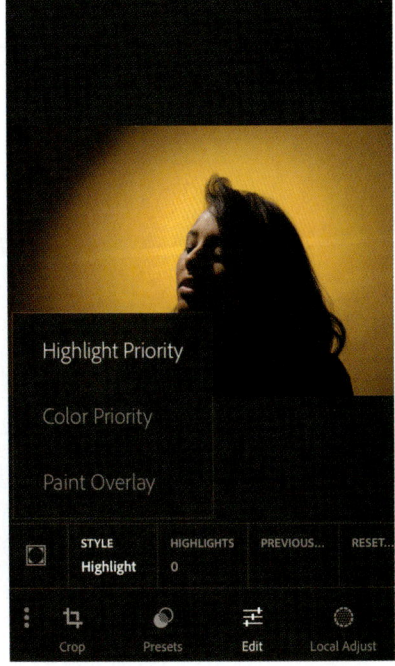

Priority and Paint Overlay. Use Highlight Priority to suppress vignetting on the highlights in your photo. Color Priority suppresses vignetting on colours around the edge of the frame. And Paint Overlay masks anything, regardless of brightness or colour, with the vignette.

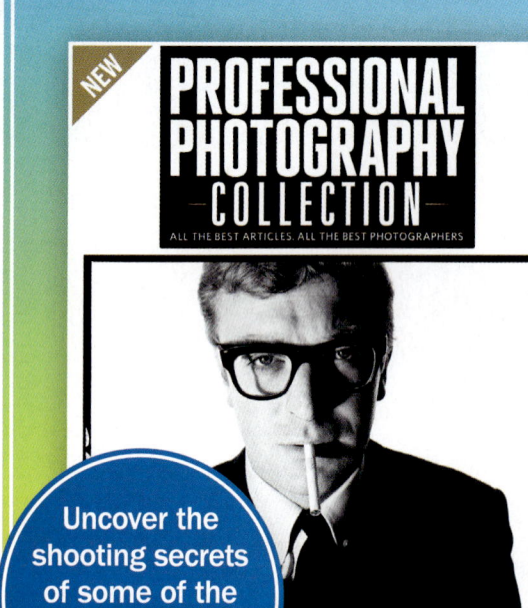

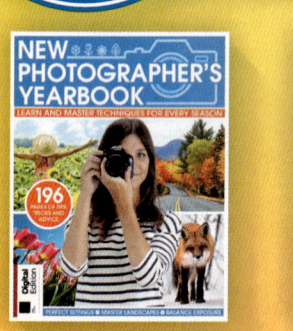

Uncover the shooting secrets of some of the world's greatest photographers

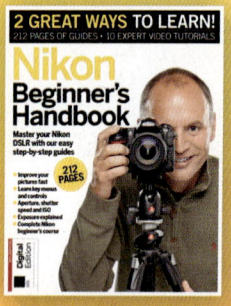

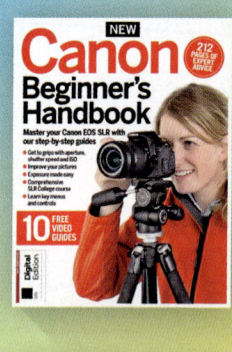

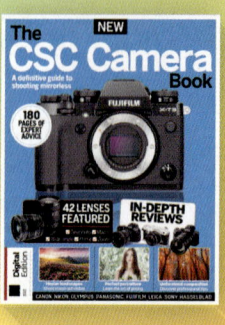

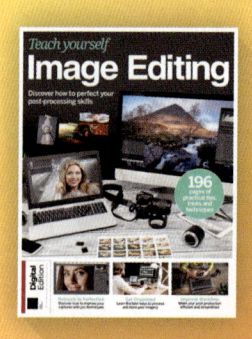

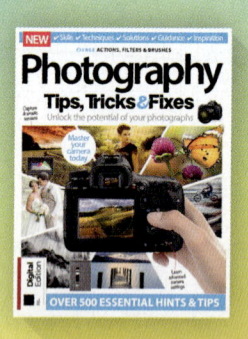

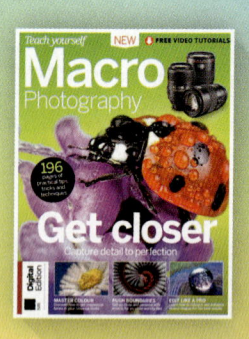

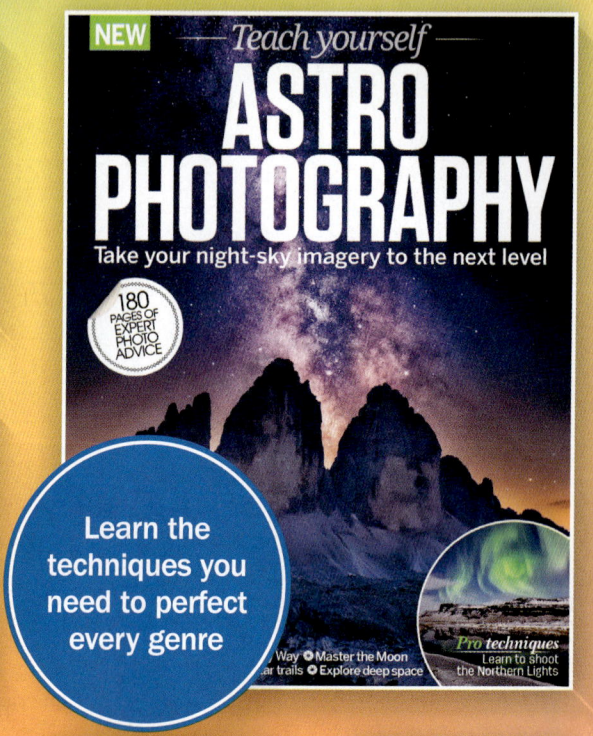

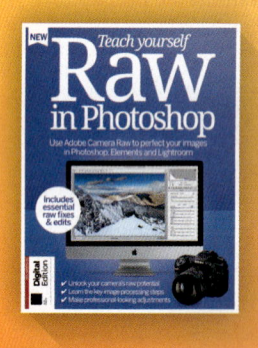

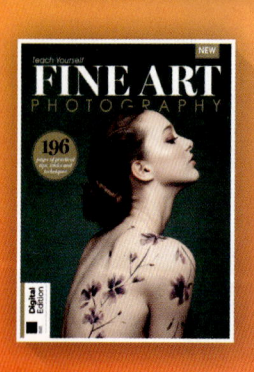

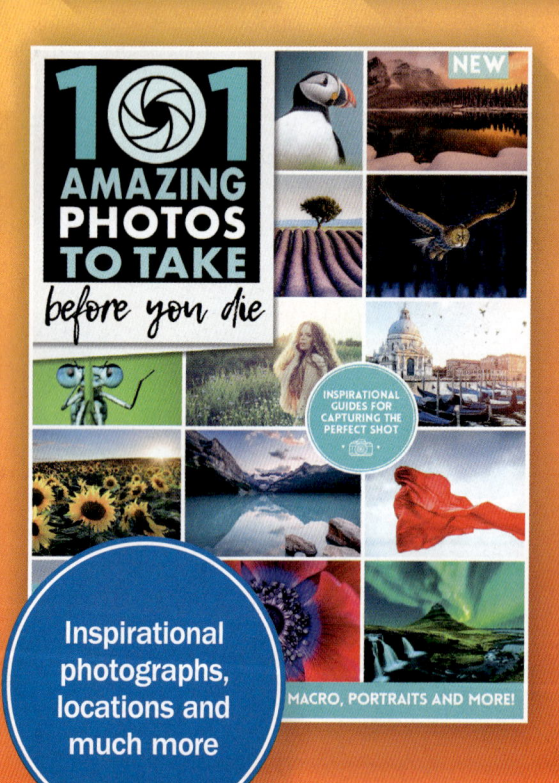

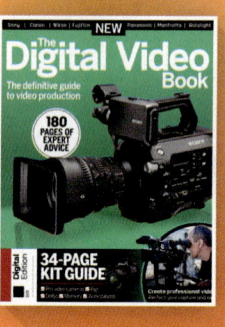

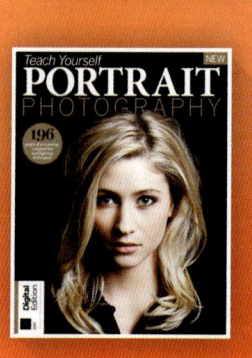

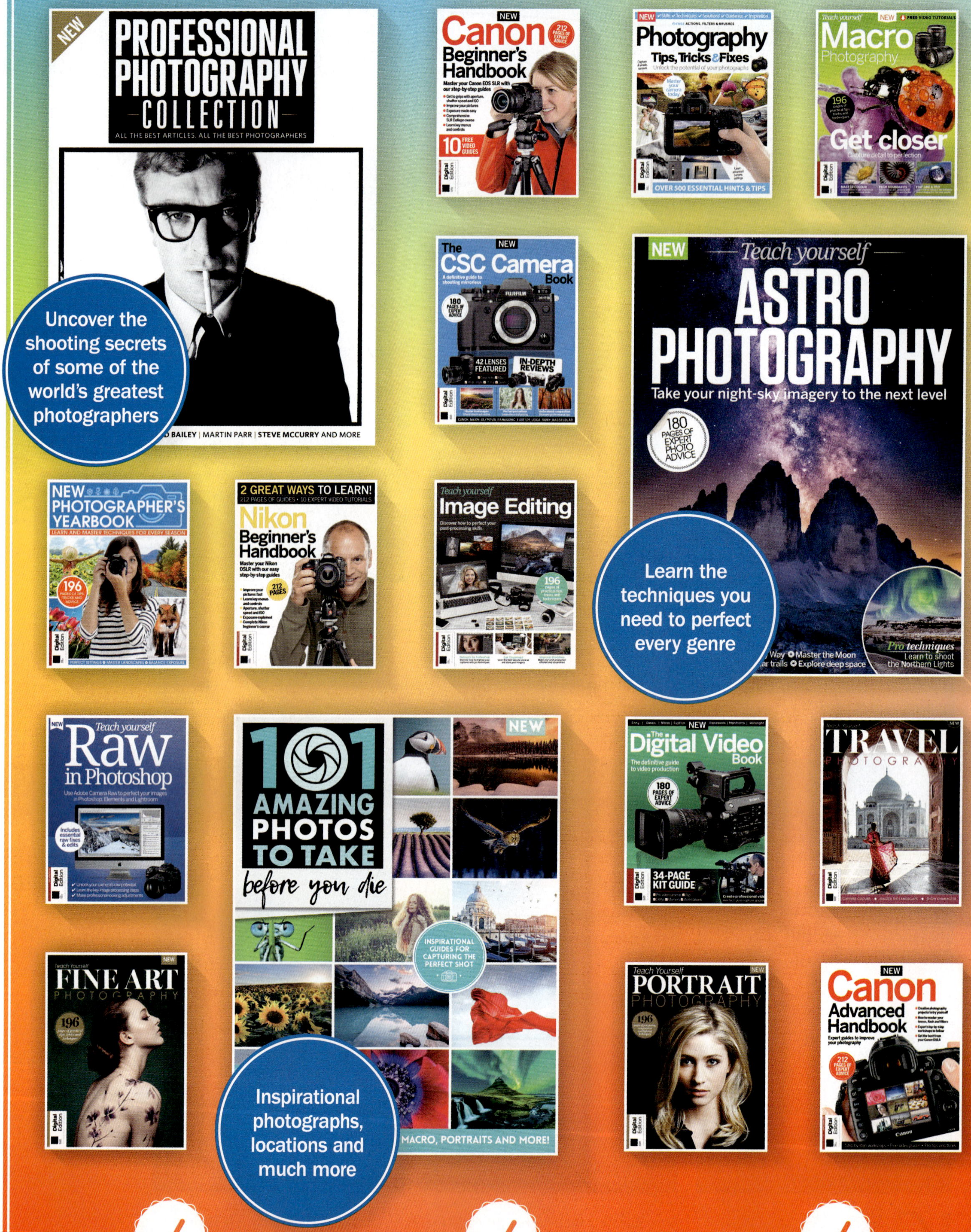

Learn the techniques you need to perfect every genre

Inspirational photographs, locations and much more

✓ Get great savings when you buy direct from us

✓ 1000s of great titles, many not available anywhere else

✓ World-wide delivery and super-safe ordering

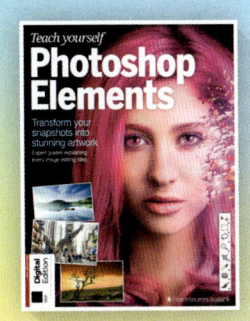

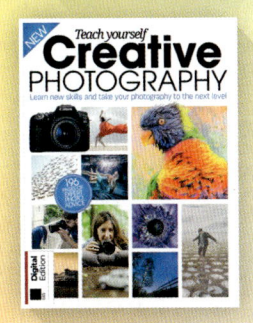

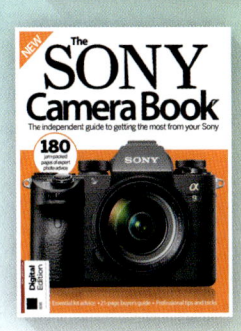

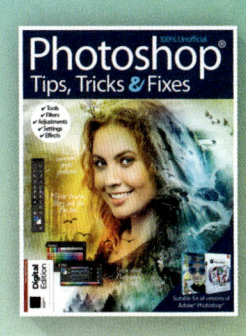

MASTER YOUR CAMERA WITH OUR BOOKAZINES

From tech to techniques, learn to take better shots with our essential photography guides and handbooks

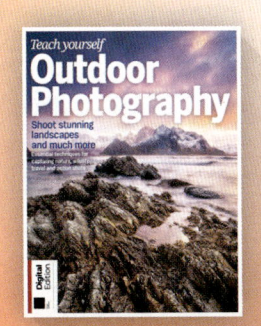

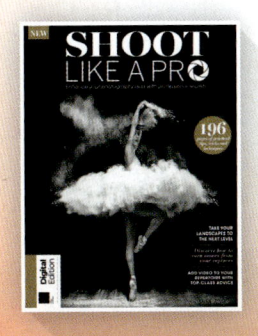

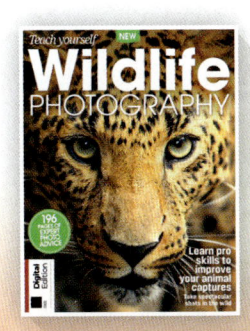

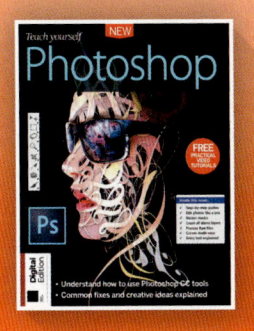

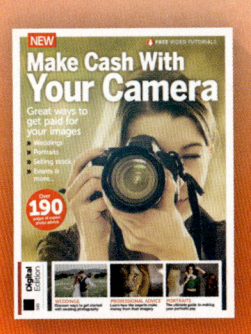

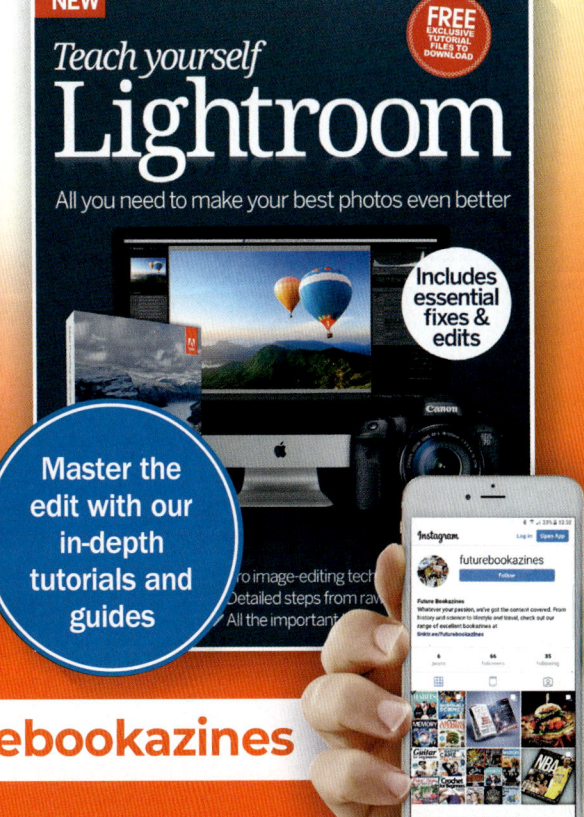

Master the edit with our in-depth tutorials and guides

Follow us on Instagram 📷 @futurebookazines

www.magazinesdirect.com

FUTURE

Magazines, back issues & bookazines.

SUBSCRIBE & SAVE UP TO 61%

Delivered direct to your door or straight to your device

Choose from over 80 magazines and make great savings off the store price!

Binders, books and back issues also available

Simply visit www.magazinesdirect.com

✓ No hidden costs   🚚 Shipping included in all prices   🌐 We deliver to over 100 countries   🔒 Secure online payment

FUTURE

magazinesdirect.com
Official Magazine Subscription Store